Karel Rogers

Thinking Green
Ethics for a Small Planet

Copyright © 2010 Karel Rogers
All rights reserved.

ISBN: 1449574955
ISBN-13: 9781449574956
Library of Congress Control Number: 2009911724

Dedicated to
those who see problems
and
feel powerless to fix them.

Table of Contents

Prologue: An Invitation · xiii

How This Book Is Organized · xiv

Section I—Putting things in perspective is a critical first step in modifying our behavior as individuals and as a species. · xiv

Section II—In order to make ethical choices we need to recognize that we are a part of the natural world and act accordingly. · xv

Section III—We face serious challenges as we seek to change our relationship to the natural world. · · · · · · · · xvi

Section IV—Freedom and prosperity are possible when we make ethical, sustainable choices. · · · · · · · · · · · · · xvii

Section I Gaining Perspective · 1

Chapter 1 Time, Species, Global · 3

Time · 3

Species · 10

Global · 13

The I, Me, Now Model · 13

Environmental Problems · 15

Time, Species, Global Perspective · · · · · · · · · · · · · · · 18

Important Concepts of this Chapter · · · · · · · · · · · · · · 25

Chapter 2 Obligate Consumption · · · · · · · · · · · · · · · · · · 27
    The Onion, Layer One: Structural Consumption · · · · · 29
    The Onion, Layer Two: Marketing and Eating · · · · · · · · 34
    The Onion, Layer Three: Quarterly Earnings · · · · · · · · 37
    The Onion, Layer Four: Planned Obsolescence · · · · · · · 39
    Life in the Onion · 41
    Important Concepts of this Chapter · · · · · · · · · · · · · · 53

Section II Rethinking Core Issues · · · · · · · · · · · · · · · · · · · 55
    Chapter 3 People Are a Part of Nature · · · · · · · · · · · · · 57
        Mice · 57
        The Evolution War · 61
        Philosophical Options · 65
        Evolution—A Simple View · · · · · · · · · · · · · · · · · · · 68
        Evolution—A Modern View · · · · · · · · · · · · · · · · · · 70
        Evolution in Humans · 72
        Species Affect Each Other's Evolution · · · · · · · · · · · · 74
        Non-biological Evolution around Us · · · · · · · · · · · · · 78
        Speedy Evolution in Culture · · · · · · · · · · · · · · · · · · 82
        Using GDP to Measure Economy Impacts
        Economic Evolution · 83
        Ethics and Science Work Together · · · · · · · · · · · · · · 88
        Important Concepts of this Chapter · · · · · · · · · · · · · · 90
    Chapter 4 All Species Modify Their Environment · · · · · · 93
        Orkin Rules · 93

The Origin of American Bad Habits · · · · · · · · · · · · · · · 98

Spreading Success and Spreading Problems · · · · · · · · · · 102

Design Goals and Unintended Consequences · · · · · · · 104

Clinging to Our Old Ways · 107

Important Concepts of this Chapter · · · · · · · · · · · · · · · 110

Chapter 5 Nature's Operational Rules · · · · · · · · · · · · · · · 113

Nature's Principles · 115

Waste Equals Food · 116

Nature Runs Off Current Solar Income · · · · · · · · · · · · 120

Nature Depends on Diversity · · · · · · · · · · · · · · · · · · · 124

Information-rich Solutions · 130

Design Rules · 136

Important Concepts of this Chapter · · · · · · · · · · · · · · · 137

Section III The Challenges We Face · · · · · · · · · · · · · · · · · · 139

Chapter 6 The Commons Concept · · · · · · · · · · · · · · · · · 141

Easter Island · 141

The Tragedy of the Commons · · · · · · · · · · · · · · · · · · 144

Commons in Our Lives · 145

Woodchuck Slobs · 148

The Tragedy of the Oceans · 154

Industrialized Fishing Technology · · · · · · · · · · · · · · · 156

Killing the Food Web · 157

Non-solutions · 160

Successful Social Arrangements · · · · · · · · · · · · · · · · · 162

Important Concepts of this Chapter · · · · · · · · · · · · · · · 167

Chapter 7 Food and Water · 169

    The War with the Newts · 169

    The People of the Corn · 172

    Soil Health Is a Commons · 176

    Fossil Water Is a Commons · 176

    Industrialized Meat-making · 179

    Beef and Global Warming · 181

    Ocean Damage and Other Unintended Consequences · · · 182

    Water Limits · 185

    Collapsing Civilizations · 188

    Local Food, Local Water, Local Collaboration · · · · · · 190

    Natural Abundance and Biodiversity · · · · · · · · · · · · · · 193

    Natural Water Abundance · 196

    Important Concepts of this Chapter · · · · · · · · · · · · · · · 200

Chapter 8 Land Use Traditions · 203

    Public Lands—Preservation and Conservation · · · · · · 204

    Private Land · 207

    Townships or Cities · 210

    Corporations · 210

    States · 212

    The American Dream · 213

    Wealthy Americans · 214

    Poor Americans · 216

    Non-human Community · 218

Pollinators · 220
Changing Our Land Use Practices · · · · · · · · · · · · · · · · 221
Compact Development · 223
Farmers as an Endangered Species · · · · · · · · · · · · · · · 225
Re-knitting Nature around Us · · · · · · · · · · · · · · · · · · 226
Re-knitting Human Community · · · · · · · · · · · · · · · · · 228
Important Concepts of this Chapter · · · · · · · · · · · · · · 233
Chapter 9 War—An Evolved Piece of Cultural Insanity · · · 235
History of War · 235
War Is a Uniquely Human Practice · · · · · · · · · · · · · · 236
The Role of Feeding Ecology · · · · · · · · · · · · · · · · · · 240
Predicting the Winners of War · · · · · · · · · · · · · · · · · 241
We Live with Our Frozen Accident · · · · · · · · · · · · · · 244
Conflict Resolution Nature's Way · · · · · · · · · · · · · · · 245
Environmental Impacts of War · · · · · · · · · · · · · · · · · 247
Only the Big Have a Say · 250
War Is an Abuse of Ethical/Moral Behavior · · · · · · · · · 252
Shifting from War-like to Peaceful · · · · · · · · · · · · · · · 253
Bringing Home the Bacon · 256
Ethical Globalization · 258
Important Concepts of this Chapter · · · · · · · · · · · · · · 261

Section IV Freedom and Prosperity · · · · · · · · · · · · · · · · · 265
Chapter 10 Making Choices in Our Personal Lives · · · · · 267
The Secret of Abundance · 269

What Matters Most to You? · 271

Trade Offs · 272

The Fly in the Ointment · 276

Simplifying Environmental Concerns · · · · · · · · · · · · · · 279

Determining the Big Things · 280

Big Ticket Item #1: Choosing Where to Live · · · · · · · · ·282

Big Ticket Item #2: Energy Efficiency · · · · · · · · · · · · 284

Big Ticket Item #3: Energy Star Appliances · · · · · · · · 289

Big Ticket Item #4: Cleaning Products · · · · · · · · · · · · 290

Big Ticket Item #5: Moral Meat · · · · · · · · · · · · · · · · · 291

Big Ticket Item #6: Local Food · · · · · · · · · · · · · · · · · 296

Big Ticket Item #7: Clothing · · · · · · · · · · · · · · · · · · · 297

Big Ticket Item #8: Buy Local · · · · · · · · · · · · · · · · · 298

Big Ticket Item #9: Recreation · · · · · · · · · · · · · · · · · 300

Purposeful Design in Life · 301

Important Concepts of this Chapter · · · · · · · · · · · · · · 303

Chapter 11 There Is No Hope—How Wrong
Can You Be? · 305

Loss of Hope · 305

Government Follows, Not Leads · · · · · · · · · · · · · · · · 308

Punctuated Change ·310

Biomimicry ·312

Cradle to Cradle ·313

Natural Capital and Profit ·317

Viral Marketing of the New Thinking · · · · · · · · · · · · ·321

Corporate Pioneers · 322
The Grid and Alternatives · 329
Energy for Vehicles · 336
Industrial Ecology · 338
No Need for Green Wash—There's More Profit
in Green · 338
Important Concepts of this Chapter · · · · · · · · · · · · · 340
Chapter 12 The Great Unfolding Story · · · · · · · · · · · · 343

Prologue: An Invitation

From the food we eat to the houses we live in to the cars we drive, we live in a bubble of unreality disconnected from our roots as humans, as a biological species, and as citizens of planet earth. I invite you to join me in a journey to expand your worldview into one steeped in the reality on which our healthy existence depends. As we take this journey, I hope you see how your everyday decisions are actually moral pivot points and also how your spiritual life can expand into a rich, enjoyable, and moral philosophy that can improve our lives today and form a firm foundation to carry our species forward successfully into geological time.

I hope this book motivates you to help transform mainstream American culture from the dirty brown of the industrial revolution to the green of a new behavioral and technological revolution for humans. I also hope this book helps you to recognize your power and responsibility as you spend your money, vote, monitor your elected representatives, make decisions at work, and parent your children. What we choose does matter, possibly more than any individual can imagine.

You may ask why I am targeting mainstream American culture in a book about global philosophy on sustainability. The reason is simply that American culture is the most powerful global influence that exists today. That may feel wrong to you but think back to the era of European colonialism. The relatively small group of people living in France, Great Britain, and Spain literally shaped life for nearly everyone on every continent and their influence is still with us today. The US exists as it does with our constitution and bill of rights because of the historical influence of European colonialism. Similarly, our modern world is being shaped by American free market capitalism and by American-dominated institutions like the World Bank, International Monetary Fund, and the World Trade Organization. Around the world, US influence shapes economies, determines export markets, and underpins currencies. The US now wields unprecedented global influence in technologies used and in the details of lives on every continent of the earth as well as in the ocean. Thus, we as Americans (including people in other countries living an American lifestyle) have unprecedented responsibility to lead the world toward a sustainable future, not just politically but also in how we conduct our personal lives.

How This Book Is Organized

This book is divided into four sections: getting perspective on this time and place; rethinking core issues in our relationship to the rest of nature; looking at the challenges we face as we seek to influence our cultural evolution; and how we can find freedom and prosperity by improving the quality of our lives while enhancing earth's life support systems.

Section I—Putting things in perspective is a critical first step in modifying our behavior as individuals and as a species.
The slice of time in which we live is tiny compared to earth's history and we are only one—*one!*—of 1.8 million

known species on the planet. As Americans, we live in a bubble of unreality designed to increase consumption. Our patterns of overconsumption damage the environment, injure our physical and psychological health and well-being, and threaten our continued existence. Fortunately we have many ways we can help renovate mainstream American culture that doesn't entail giving up electricity and living a life of deprivation.

Section II—In order to make ethical choices we need to recognize that we are a part of the natural world and act accordingly.

Although humans and all of our activities are a part of nature, we have let our religious traditions and scientific/technological advances lead us to believe we are separate and different from the natural world. In fact, because humans are inextricably bound to our home on earth, we are evolving right along with the rest of nature. The traditional environmentalist assumption that humans must destroy to prosper is wrong. But we can't go on indiscriminately undermining the foundations of the ecosystems that form earth's life support systems we depend upon. All species modify their environment—the key is to redesign our activities to fit those that promote human well-being in the context of the time-tested constraints that non-human species live by—essentially to re-knit the living world back into our everyday lives. Finally, our society and its members cannot make truly ethical choices if our primary decisions are based only on cost and the economy. What superficially appears cheapest is not always best for us or for the rest of nature, and some things shouldn't be purchased at all even if they appear cheap because they destroy earth's life support systems.

Section III—We face serious challenges as we seek to change our relationship to the natural world.

First, we all share "commons"—resources like the oceans, air, biodiversity, and quiet and darkness that we all need and have a right to but don't have to pay for. Decades of careless use have damaged some of these commons, and others are in danger of collapsing. Globally, the most immediate and dangerous result of a failing commons is the loss of productivity of earth's oceans. Because fishing has become a technologically-driven enterprise of scooping whole populations of species out of our oceans while killing all of the "bycatch," the complex food webs of the seas are crumbling. The primacy of economic considerations in our social decision making is another challenge we face and another example of a commons. Free markets of all types are a commons whose rules of profit can be determined by powerful interest groups so the spread of destructive production and consumption patterns for basic human necessities—food and water—has become a global disaster. We are also challenged by our history. A case in point is the land use traditions we use in the United States and the consequences of those traditions on ourselves and the rest of nature. We are further challenged by our deep evolutionary history, in particular our acceptance and support for war and war-like behavior. Conflict resolution is common in our non-human neighbors, but killing in wars is restricted to humans and their nearest kin, the chimpanzees—a legacy of our evolutionary past. But surely a species of our intelligence should be able to develop more benign methods of conflict avoidance and conflict resolution. Central to all of these challenges is the fact that we live embedded in a series of evolving systems so each choice we make, from our food to how we measure our national prosperity, puts a selective force on many levels of biological, cultural, and global well being. Collectively, we humans and our activities have recently become a force of nature. Thus, we have a deep ethical responsibility to choose our personal, social, and economic goals carefully.

Section IV—Freedom and prosperity are possible when we make ethical, sustainable choices.
To prosper in an ethical way as individuals and as a species we must free ourselves from wasteful consumption and avoid consumption that doesn't improve our quality of life. By carefully choosing where we live, what we live in, what we do inside our homes, and what we do for recreation, we gain the benefits of free time, health, and a rich spiritual life. As we make these considered and responsible choices, the rest of living things near us and in other places on the globe will also prosper. It's reasonable to ask how only one person can make a difference that reverberates around the globe to transform human destiny. The simple answer is because you wouldn't be alone. Internationally, there is a growing movement of people who recognize the importance of this moment in time as an opportunity for a better future.

* * *

There's nothing easy about what I'm advocating; this isn't a 100-easy-ways-to-save-the-earth kind of book. Instead, I am advocating a different way to look at the world so our choices allow us to have hope for the trends of the future. This means all sorts of things must change. But as those around us see how we can prosper economically, impact our communities, and grow spiritually even with things like the global financial meltdown occurring around us, then more people will make decisions that will avert a truly threatening global ecological and cultural meltdown. With the background you gain from this book, you will be able to discern ethical choices from the cacophony of green wash spewed out around us each day. Through the actions of local individuals acting in right ways, the great unfolding, evolving story of the universe will continue to include humans. That, I think, is a future worthy of our very special species.

Section I

Gaining Perspective

Chapter 1

Time, Species, Global

My first goal is to stretch your perception of what is important in three major dimensions: *time*—cosmological, geological, historical, and futuristic time; *species*—tiny to large, ugly to charismatic, and ordinary to extraordinary; and *global*—from your home to your town to your country to earth's oceans to the smallest island nation on earth. Shorthand for this is what I call a "time, species, global" perspective. Once we have this perspective, we can begin to make decisions that will heal ourselves and our planet and redeem our future.

It is a testament to the biological success of our species that humans have the power to choose a destiny for ourselves and for the species around us. In everyday life, humans are a ubiquitous and overpowering presence affecting every square inch of earth and much of the space around it. That makes it easy to forget how dependent we all are on Earth's life support systems and each other. It also makes it easy to forget how small a speck we represent in the great cosmic story of the universe. Hearing that story should expand our time, species, global perspective.

Time

Humans tend to think of time in months, days, and years, all measurements related to a human lifetime. Yet the things that impact us occur over billions of years down to fractions of a second. Our universe is 13,700,000,000 years old and contains hundreds of billions of galaxies, all equally old and all originating in the Big Bang (or maybe the Big Bounce). Subatomic particles, atoms, galaxies, stars, and planets all originated one after the other in response to local conditions and chance events. Our galaxy, the Milky Way, has over 100 billion stars. Earth's sun is an average star of the Milky Way Galaxy, and Earth is one of nine planets orbiting our sun. Sounds like a lot, doesn't it? Yet visible matter makes up only 4% of the known universe—the rest is 22% dark matter and 74% dark energy, both of which we know exist but can't see. As a matter of fact, dark matter and dark energy go straight through everything we can see, including our own bodies, as though the visible world doesn't exist. When you take all this into account, I hope you notice how small our place in this universe is.

Earth itself is 4.56 billion years old. On the path to our modern world, about ¼ of Earth's history was taken up with the evolution of molecules. This huge piece of time is essentially a story of the accumulation of the ingredients for life, some of which may have originated on Mars or elsewhere in the Milky Way. Regardless, eventually there was a diverse community of entities we might recognize as being alive—essentially the first cells. These early cells were a community with much less rigid rules for living than we have today—they died easily and their components were easily taken up and used by other cells, processes we still see occurring among bacteria today. Possibly it's best to think of this time as communal sharing of innovations. Functional breakthroughs that increased organization, utilized available energy sources to decrease disorder in the cells' interiors, and mechanisms that improved the ability of successful cells to replicate were propagated widely and "stolen" by other cells.

About 2/3 of the way through Earth's history, breakthroughs occurred whereby cells became much more complex, not by the development of a whole new type of cell but rather by a close collaboration between several types that had previously existed independently. One way to visualize this breakthrough is to think of the early cells as a tiny RV. They contained everything necessary for living but it was all bare-bones: the table had to be folded up to allow the bed to fold down and the floor of the toilet did double duty as the floor of a shower. The new cells were like the mansions of today with a separate room for everything: several kitchens each with a variety of ways to cook food, a sun room for breakfast, a porch table for lunch, and a formal dining room for dinner. Our cells are eukaryotic (true cells), mansions whose origin is from ancient collaborations that house previously free-living organisms as cellular organelles. One important example is the cellular organelle called the mitochondrion. Mitochondria still have their own kitchen (genes and the equipment to express their genes) but they now produce energy for the whole cell. They are important to us because the energy needs of organisms like ourselves cannot be met by the methods of the early tiny RV cells. We can be grateful our early ancestors were cooperators! Without their cooperation, the next break though could not have happened.

Then, about ¾ of the way through Earth's history, some eukaryotic cells developed the ability to live in tightly knit, interdependent, synergistic communities that today we call multicellular organisms (animals and plants). This innovation was also the origin of embryology, a process dependent on a highly conserved genetic program that can take an egg from one individual and a sperm from another to develop an offspring that has DNA from both parents. It is slight variations in the embryological program that allows the myriad of physical forms of plants and animals we see today on Earth.

The story from here forward gets more exciting because we can see more of ourselves in it. With about 5/6th (83%) of its history behind it, Earth entered the age of fish—at the beginning of this age there was no life on land but before its end, millions of years of biological terrestrial productivity had stored ancient sun energy in the form of the coal and oil we use today to power our lives. The age of fish ended with "The Great Dying," catastrophic ecosystem collapses caused by massive climate changes. About 250,000,000 years ago, the age of dinosaurs started but it also ended abruptly by the impact of a giant meteor that hit the earth about 65,000,000 years ago. This was a "frozen accident", a chance event that could have happened differently (or not at all). Frozen accidents are pivot points that determine subsequent events.

The vulnerability of the giant reptiles to the catastrophic ecological collapse caused by the meteor's dust cloud and disruption of weather and food was a boon for mammals. Mammals had been around during most of the age of dinosaurs, but with all the ecosystems dominated by giant reptiles, there wasn't a lot of ecological opportunity for mammals to diversify. They lived furtive lives hiding in burrows coming out mainly at night to forage while the reptiles were more sluggish. Because of our remote ancestors' fringe life-style, some were able to live through the catastrophic land and ocean ecological collapses that occurred as a result of the meteor, a frozen accident that worked to our benefit.

It takes a while for ecological communities to recover from catastrophic collapses of ecosystems because everything from the balance of oxygen in the atmosphere to water availability is pretty much determined by the collective activities of the species living together in mature ecosystems. Still, mammals like ourselves who use a placenta to nurture their embryos within the female's uterus survived, mutated, and radiated to fill the niches left open by the vanished reptiles.

It was only about 2,000,000 years ago—a mere .014% of earth's history—that the genus *Homo* evolved among a diverse group of great apes that lived in Africa. Our ancestors then spent most of those two million years developing a large brain and fine tuning our embryological blueprint to wire our brain cells so we can do things like speak to one another and understand one another's intentions. As our bodies and abilities improved, so did our technology. Even though our brains were already quite large, we spent most of the ice ages using stereotypical massive stone tools and living as hunter-gatherers. But once the wiring in our brains became modern, we moved quickly through the middle and new stone ages and then through the iron and bronze ages into agriculture, towns, and into the history we've been taught in school.

It's only been during the past 200 years or so that we have harnessed the ancient sun's energy stored during the age of fish to do work for us. These 200 years of the Industrial Revolution encompass only 0.000001% of Earth's history, a tiny fraction of time. The key elements of U.S. consumption patterns have developed mainly since WWII, a period encompassing only about 1/4 of the Industrial Revolution. Therefore, the "reality" we've created using fossil fuel with our millions of cars, air-conditioned homes, and vast tracts of turf grass cannot be a reality that fits within Earth's evolved systems or into the reality we humans lived with during most of our existence. And certainly, the point of cosmic history cannot be those of us living today with particular religious persuasions. We are too small a part of what exists to be so centrally important to all of existence. If we were, at a minimum I would expect at least dark matter and dark energy would give us some special treatment.

This point becomes even more pronounced if one considers the multitude of species that live on the earth today—there are 1.8 million named and known multicellular species with which we share the earth. That's not counting the countless species that have lived in the past, and it's also not counting the ubiquitous microbes that live out their lives in the water, soil, rocks, hot springs, and ocean depths. With innovations in DNA technology, scientists are just now starting to identify the modern community of bacteria-like organisms that have inhabited the earth for ¾ of its history. So now we have it down to humans as one species out of more than 1.8 million living on one planet of nine orbiting one sun out of billions located in one galaxy out of billions in this visible universe, which encompasses only 4% of matter and energy that exists.

I hope I'm making myself clear. We have lived by an "American-centric" ethos for only 0.000001% of the Universe's history (0.000003% of Earth's history). This means we've been ignoring about 99.999999% of the history of how our home works! Is it any wonder we have the horrible social, environmental, and economic problems we have today? Is it any wonder specialists of all sorts are warning us daily of impending disasters? Is it any wonder that people like Thomas Friedman, a frequent global traveler and columnist for the *New York Times*, are writing books like his 2008 *Hot, Flat, and Crowded: Why We Need a Green Revolution—And How It Can Renew America*? Other examples are Fred Krupp and Miriam Horn's *Earth: The Sequel: The Race to Reinvent Energy and Stop Global Warming* and James Speth's *The Bridge at the Edge of the World: Capitalism, the Environment, and Crossing from Crisis to Sustainability*. Their detailed messages differ, but their themes are the same: It is time to wake up—we must change our behavior now to avoid catastrophe.

I agree. The ethical basis for our decisions in our lives, careers, consumption, spirituality, and citizenship from the local to the global level must be informed by considerations of where we fit in the greater scheme of things. These choices must also be based on the understanding that this universe and everything in it proceeds as a great evolving system where actions do have impacts far beyond their immediate vicinity. Each action, each decision we make is a vote in the great, evolving story that has been unfolding here on Earth.

In the past 150 years, and especially since WWII, we haven't been doing a very good job of using our votes wisely. We've put greed, passivity, and ignorance in the driver's seat and with our wealth have taken the globe on Walt Disney's "Mr. Toad's Wild Ride." Thomas Friedman refers to this as "dumb as you wanna be." What an apt description of lives devoid of global purpose, deep spirituality, and worthwhile goals but glutted with entertainment, luxuries, and irresponsibility! We don't have to live this way—we can improve the quality of our physical lives and deepen the quality of our spiritual lives if we are willing to wake up and make truly ethical decisions, decisions based on an unassuming time, species, global perspective.

We humans around the globe have some serious work to do if our species is to prosper in the future. Each of us has an important role to play—that includes CEOs, government officials, small businesspeople, and the common person with a pocket book and a vote.

But we cannot look to government to take care of our problems for us and certainly huge corporations won't do it. It's up to each of us as a citizen of earth to take up the challenge and to become part of a movement that changes the trajectory of disaster we're on. This means making ethical personal choices and demanding our government behaves in an ethical way also. So far we haven't done that. Civilization-threatening problems like global warming get a superficial brief mention and then we re-focus on what's important to us—the economy, Paris Hilton, or the Lions. Polish satirist Stanislav Lec said, "It's true that we're on the wrong track, but we're compensating by accelerating."[1] Like a train with five engines, we're accelerating on a track laid down in the past two hundred years of our history. Those 200 years represent only a fraction of a second in the history of life on earth and yet we're behaving as though this world of our making is somehow normal and rational. All around us are signals screaming messages of disaster at us and yet we blissfully speed ahead faster and faster.

Species

The last thing Alex said to Irene the night he died was "You be good. I love you. See you tomorrow." He wasn't very old, only 31 when he died, and his death was unexpected—apparently a heart arrhythmia coupled with a slight case of arteriosclerosis. By the time of his death, Alex could count up to eight, knew five shapes, seven colors, and about 50 word ideas. He understood the concepts of same/different, bigger/smaller, and the abstract concept of zero. He put these ideas into sentences to make his

---

[1] Herman Scheer, "Energy Autonomy: The Economic, Social, and Technological Case for Renewable Energy," quote from Polish satirist Stanislav Lec, 2007

desires known and to communicate with Irene. At his death, the *New York Times* published an obituary. Most would think Alex was a mentally impaired person, but he was actually an African grey parrot who worked with Irene Pepperberg, a professor and researcher at Brandeis University and Harvard.

Alex wasn't alone in his ability to communicate with humans in our language—other researchers have worked with elephants who are conscious of themselves, lemurs who can count and order sequences, sheep who recognize individual human faces and remember them long term, and Betsy, a border collie with a vocabulary over 340 words[2]. These are just a few examples of animal minds around us. Their lives and their species' existence are intricately tied with humans. The amazing thing is that uniformly these animals are communicating with us in our terms but we aren't communicating with them in their own context/language/signals. And yet we have traditionally looked at animals as though they were trainable machines without self-awareness. How many of us take the non-human minds around us into consideration when we make decisions that affect them, their families, and their homes?

We tend to think of our own species as the center of everything that is important. Well, actually, maybe we think of a fraction of humans as most important, those who are rich and famous as though being rich and famous makes you somehow more important than other people. The reality is that as normal humans we have over 6,000 species living in, on, and around our bodies, hundreds in our gut alone—these species are vital to keeping us healthy. And those are only our closest

---

[2] "Minds of their own: Animals are smarter than you think", National Geographic, March 2008.

neighbors—remember, there are 1.8 million named multicellular species besides ourselves.

Our judgment about the right or wrong of our actions depends on the extent of our circle of concern during decision making. That circle of concern doesn't always include many other people, much less other species. My purpose in this book is to increase the scope and flexibility of your circle of concern so your worldview takes into account cosmic history, the evolving community living on Earth, and the long-term sustainability of humans as a biological species in addition to our current individual welfare and responsibilities. If the only thing of importance in our worldview is our personal desires and the welfare of our families and friends, then our ethical decisions turn into millions of short-term "good/moral/ethical" choices that add up to a massive disaster for humans and the rest of life on this planet. This challenge to broaden our worldview is not trivial but it fosters a richer spiritual life because connecting deeply to this place we live forms a foundation for seeing our home with new eyes.

Humans are inextricably linked to our heritage as biological organisms that evolved in the context of nature's operational rules. That is, in our long history we lived on current solar income, we prospered because of the diversity present in ourselves and in the non-human species around us, our waste became the food for other species, and we lived with an acute awareness of our dependence on the natural world. Now, with our artificial separation as creatures distinct and above the rest of nature, we face serious challenges to re-knit our lives back into a design that respects and mimics the time-tested rules of our home on earth. The spiritual richness that comes from re-connecting with our biological heritage has tremendous repercussions in our ability to feel the presence of power beyond ourselves.

Global

And finally, it's important that we add a global dimension to our thinking. In recent years, globalization is a common theme, but that theme focuses on movement of money, goods, materials, and jobs to various countries. It doesn't take into account an awareness of how we are crashing ecosystems by moving species and diseases haphazardly around the globe nor does it take into account how a wealthy country like ours can buy our way out of problems temporarily by off-loading those problems on the poor, uneducated, and unwary. Lastly, it doesn't take into account the cost of extracting resources from the societies, ecosystems, and environments in far reaches of the globe nor the cost of wars to ensure a steady flow of resources from other countries.

So when we consider a global perspective we're not thinking of economic and cultural globalization—the movement of money, goods, raw materials, and invasive species from country to county—but the practice of considering how these activities affect the health and long-term welfare of ourselves, the other species of the earth, and the earth itself. If we use earth's model of globalization rather than our current industrial model, we can protect the prosperity of the future while promoting our own prosperity. We can't possibly make good ethical decisions without deeply understanding this place and expanding our thinking to include an understanding of the vital heritage we've received from the diversity present on earth.

The I, Me, Now Model

People immersed in mainstream American culture base decisions on the corporate quarterly report model—essentially a three-month window either side of now that is measuring nothing more than profit for one corporation. Translated to our personal lives, this means we buy and assume debt until we are limited by how much we can pay per month—never mind perturbations in the system like an interruption in our paycheck or an unexpected increase in the cost of heating oil or food. This "just-on-time" mentality is a decision-making perspective based on I, me, and now rather than time, species, and global. I, me, and now gets us in all sorts of problems.

Take the American car industry, for example. In November 2008 the CEOs of the big three American automakers flew to Washington in their private jets asking for a bailout because the recent global financial meltdown caught them flat-footed. If they go bankrupt, the repercussions through the global economy will severely harm countless people directly or indirectly employed by them. The management teams of these businesses have focused on selling bigger and fancier vehicles, manipulating federal regulations to avoid fuel efficiency, and investing in campaigns to fool the public into questioning the science of global warming, all because their profit margins were higher on big vehicles. Now, with gas prices periodically hitting $4 per gallon and promising to go higher, the big three's gas guzzlers have poor resale value and very few people are willing to buy a new SUV, even if they are practically given away. The big three's executives have been aware for at least 15 years that this day was coming and yet they ignored the signals. A perspective of I, me, and now has not served their companies well in the long term and their narrow perspective has also created a problem for all of us.

With the help of the auto industry, Americans now face a choice between losing a key piece of their economy and bailing out an industry that has manipulated markets for short-term gain and ignored the damage their products are doing to the global climate system. Individual people who bought into the carmakers' vision are now stuck with huge car payments on gas-guzzling, roll-over-prone SUVs. Anybody with a perspective that accounts for anything beyond quarterly profits must judge the automakers' decisions over the past 15 years as highly unethical. And those who believed the automakers' ads on TV—zero percent interest, rebates, and other sales gimmicks—are now paying a steep price.

The car industry is just one example out of thousands that could be used. Taken together, the cumulative unethical I, me, now decisions of countless "deciders" have brought humans to the brink of endangering ourselves as a species. The key case in point is the environmental problems we face today and the guarantee that, if we don't change our ways, these problems will disrupt the life support systems that underpin our civilization and harm most of us as global economic development brings billions out of poverty and into the American way of life. These aren't issues our grandchildren will have to worry about. The pace of change is such that these are problems we have to worry about now.

Environmental Problems

Environmental problems scare me to death but rather than wallow in them for an entire book, let's take a brief systematic look at them. There are three main categories of environmental concerns—human population growth, resource depletion, and pollution. As Paul Ehrlich and David Brower[3] so effectively

---

3 Paul Ehrlich and David Brower. 1970. "The Population Bomb". Ballantine.

showed us, a few affluent people using polluting technologies that destroy lots of resources have a larger environmental impact than a much bigger group of poor people using benign technologies with renewable resources. Shorthand for this is the IPAT equation:

Impact = Population × Affluence × Technology.

Technologies designed without a time, species, global perspective impact both resource depletion and pollution. New materials are taken from the rich store of resources on earth, used inefficiently to generate consumer products, and various sorts of pollution are generated in the process. For example, much of our current electricity comes from coal. Mountaintops in West Virginia are removed to mine coal, the coal is burned in power plants that waste over 70% of the energy present in the coal, and mercury and other noxious compounds are released into the air and water as pollutants. Meanwhile we Americans demand cheap electricity and use it wastefully; think about the myriad appliances in your home that display a digital clock even when the appliance is turned off.

Population growth is important in the equation because when population growth is coupled with wasteful affluence, the greater the harm to every living thing on earth. With our current model of globalization, population growth isn't limited to biological rates of reproduction; it's magnified by cultural rates of adoption of the American lifestyle. The IPAT message is written large and small in the myriad of environmental problems that exist on earth today.

It is depressing to think about these sorts of problems, but it's important to be aware of how long the list is and how the problems are escalating toward a crisis. It's also important to recognize the difference between the life-threatening, big problems and the eye-catching I, me, now problems that are used to raise funds for environmental organizations. The major environmental problems are due to the interaction of the three main categories of environmental concerns. They manifest themselves as global warming, acid rain, peak oil, loss of soil, loss of biodiversity, dirty air, urbanization, deforestation, desertification, loss of wetlands, ocean overfishing, loss of the ozone shield, species extinction, dirty water, noise pollution, light pollution, smell pollution, water shortages, drought, flooding, poor health of humans, poor health of animals and plants, spread of disease, inhumane treatment of animals, racism (the poor bear the brunt), invasive species, loss of habitat, and many more. The drivers that magnify and expand the scope of these environmental problems are bad technology, overconsumption, waste, population pressures, value systems, greed, economic pressures, political power, special interest groups, unregulated free enterprise, money, culture, ignorance, scientific uncertainty, loss of hope, exponential growth and decay, history, and false goals.

Environmental catastrophes of all sorts are horror stories on the verge of explosion now because as you read this millions of people around the globe are quickly taking on an American lifestyle through the economic benefits of globalization. More are now buying cars, eating a higher proportion of meat and vegetables in their previously rice-centered diets, and moving from their subsistence-farm villages into condos in the cities to work in manufacturing of consumer goods—things Americans have taken for granted for years. This shift hits all aspects of the IPAT equation. More people are quickly becoming affluent enough to adopt polluting, wasteful technologies that can quickly burn through the non-renewable resources we have inherited as citizens of Earth. Certainly we have no moral standing to impede other people's transition to our lifestyle but the simple truth is that the Earth cannot sustain the demands the American life style has placed on it in the past let alone the additional demands of the "new Americans."

Time, Species, Global Perspective

In contrast to the results of the I, me, now perspective, an expansive worldview—a worldview predicated on a time, species, global perspective—leads to a realistic acknowledgement of humans' small place in the great unfolding cosmic story of our universe. It also leads to a deep sense of ethical responsibility that recognizes our unique place in this cosmos as the observer and chronicle-maker of this cosmic story. As our worldview expands, our spiritual goal expands from personal goals to include recognition of our participation as active agents in the story of health and abundance and periodic catastrophe that is the natural order of earth. We come to understand we live in a place where catastrophes happen, but also a place empowered to renew itself after each death. A realistic view of ourselves as a part of nature and as an active agent of change transitions us from the lonely and isolated view that humans and their activities are separate and above nature to a deep feeling of connection to our earthly home. Bugs and worms and weeds become our neighbors rather than insignificant objects whose role in our lives doesn't count except when they irritate us with their presence. Connecting deeply to this place leads us ultimately to a much fuller spiritual life and to a healthy and abundant future for our great great grandchildren. The need for this shift in perspective and spiritual connection is illustrated by the following story.

A number of years ago I was teaching a section of a university course called Environmental Ethics. This course promoted communications skills in addition to content expectations so I had assigned a series of short papers and then a final persuasive ethics project that was to be presented as a formal paper and as a 15-minute oral presentation. To keep students from choosing narrow, trivial topics, I had developed a list of topics they could sign up for, one of which was "The Ethics of Turf Grass."

To my dismay, Mike wrote a long paper on how wonderful, functional, and ethical it is to maintain vast stretches of weed-free, insect-free, irrigated turf grass because it is so pleasant for humans in so many ways. Mike's paper fit beautifully inside the frame of modern America where acres of bright green grassy lawns are the norm, even in desert regions. This frame assumes it is a safe strategy to maintain ignorance about the impact of biologically active chemicals on the health of our children, about the limits and abundance of water, and about the finite ability of nature to provide resources and remove wastes. It also assumes it is safe for humans to maintain a lack of interest in the intricate web of species living in and above the soil, to assume a source of food and other resources from places other than "my yard," and to count on sufficient excess wealth to have the luxury of wasting the natural abundance of every square inch of earth. The sole measuring stick for the worldview Mike used was attractiveness to people who had never questioned why they should spend time, money, and effort to grow what is essentially a polluting ecological desert over vast tracts of U.S. land. In short, the paper lacked every attribute of a time, species, global outlook. How spiritually lonely, egocentric, and self-destructive to be so disconnected from the reality around us!

Fortunately Mike's error was caught on his first draft so his final paper contained exceptional insight into "The Ethics of TurfGrass" but Mike isn't alone in his original thinking. As a society we haven't been doing a very good job of nurturing time, species, global values in our youth.

You know how mothers are—they always have something to tell you about what you should be doing or seeing or reading. I'm sure they do it because they care about their kids and want to help them succeed. My mother is no different. One day she taped a show about Dr. Ben Carson, a world renowned brain surgeon who earned his fame by separating Siamese twins, and insisted I watch it. Ben Carson was from humble origins—he was the child of a divorced single mother, black, and from Detroit. When his fifth grade class had a big discussion about whether Ben Carson was the dumbest person in the world, he was willing to concede that he was the dumbest person in his class, but not in the whole world.

While Ben Carson was still in grade school, his mother insisted they turn off the TV and she required her two sons to get books from the library and write a book report on each. She sat and read each report carefully, making marks here and there on the paper. This is all the more amazing in that she was unable to read. But she knew what was valuable and lasting. The books opened a new world for Ben Carson, and he is now a force of change for young people everywhere.

Dr. Carson talks about the many young people he's met who can respond correctly to all sorts of questions about popular culture—they all know who Paris Hilton or O.J. Simpson are and they are aware of which sports teams won most recently and they know the TV schedule by heart—but who have only vague notions of anything of substance like DNA, global warming, or who is Secretary of State. He considers their knowledge and perspective to be "astonishingly superficial." In his presentation, Dr. Carson then goes on to talk about some of the defunct civilizations of the past and how the people became enamored with "sports, entertainment, and the lives of the rich and famous." Clearly, if we want our civilization to persist for any length of time, it's in our best interest to make sure our young people have a true perspective on the challenges of today and to put the pacifiers fed to us by TV[4] into their proper perspective. Sports, entertainment, and the lives of the rich and famous will not help our children deal with a globalized world nor will it help them deal with the cultural, environmental, and economic challenges of our time.

As I write this in the Fall of 2008, country after country has gone through an economic meltdown started in the US that has evaporated trillions of dollars of pseudo-wealth created by money-making schemes that have been taking place at all levels of our global society. Most of the people involved seem to be shocked that the global financial system could be so vulnerable to such huge losses. Governments around the globe are pouring manufactured money into the system in attempts to stabilize it but there is a fundamental weakness at the foundation. There are too few people doing productive work that *earns* money and too many people manipulating systems to *make* money. Time-tested values like living within our means and giving an honest day's labor for reasonable pay have been lost to instant gratification schemes that give excessive rewards for easy effort. The I, me, now perspective we use to judge ethical behavior has brought us this pain. Fortunately we have a choice in whether we keep this disastrous party going or not.

---

4   Al Gore, "The Assault on Reason", 2007, The Penguin Press

Those who hear warning messages of economic and environmental disaster pose solutions like forming small survival colonies that somehow will become islands of health and wellbeing in the future. Those people apparently haven't read about reconstructions of civilizations as they collapsed. If our social order falls apart, the strong will take from the weak, the warlike will overpower the peaceful, and few of the children will survive. That's the story we find in the archeological record—there's no reason to expect our civilization is somehow different. There's also no reason to stay on our current path and there's even less reason to think the only way to ensure our civilization will last millennia rather than a few hundred years is to tear apart our standard of living. Instead, we must take off our "dumb as you wanna be" glasses, take a hard look at the world and our lives, put our intelligence to work on the challenges we face, and activate our personal and political will to fix our problems in the context of our ethical obligations and in the context of our current scientific knowledge. Yes, this means we must change many things, but change isn't a bad thing. As Paul Romer says,[5] "A crisis is a terrible thing to waste." Let's not waste this crisis.

There's a wonderful little book that made the corporate circuit a few years ago called *Who Moved My Cheese?*[6] It's about two people, Hem and Haw, and two mice, Sniff and Scurry, all of whom live in a "Maze" and eat "Cheese." They all have little tennis shoes and each has a unique personality roughly comparable to how people react to changes around them. As

---

5 Quoted by Thomas L. Friedman, "The World is Flat: A brief history of the twenty-first century," 2005, Farrar, Straus and Giroux

6 Spencer Johnson, M.D. "Who Moved My Cheese?" , 1998, Putnam

each moves through the "Maze" and learns, there are "messages on the wall" that are demonstrated by the unfolding parable. After very rough times, Hem and Haw finally find a huge pile of cheese, so they hang up their tennis shoes and settle down in that spot in the "Maze" with guaranteed cheese, but also cheese that gets progressively staler. Meanwhile, Sniff and Scurry keep hunting for new cheese, sniffing each new find to make sure it is fresh. Haw is the most adaptable of the characters, and at the end of the book he finds handwriting on the wall that gives the following advice. "Change Happens (They Keep Moving The Cheese), Anticipate Change (Get Ready For The Cheese To Move), Monitor Change (Smell The Cheese Often So You Know When It Is Getting Old), Adapt To Change Quickly (The Quicker You Let Go Of Old Cheese, The Sooner You Can Enjoy New Cheese), Change (Move With The cheese), Enjoy Change! (Savor The Adventure And Enjoy The Taste Of New Cheese!), Be Ready To Change Quickly And Enjoy It Again & Again (They Keep Moving The Cheese)."

I love fiction because there's frequently so much deep truth in it. In *One Door Away From Heaven*, author Dean Koontz's Aunt Geneva is talking to her nephew, who must make a major life change that isn't of his choosing. She describes our situation perfectly. "Change isn't easy, Micky. Changing the way you live means changing how you think. Changing how you think means changing what you believe about life. That's hard, sweetie. When we make our own misery, we sometimes cling to it even when we want so bad to change, because the misery is something we know. The misery is comfortable."[7] This is where we Americans are right now—we're miserable but afraid

---

7 Aunt Geneva in "One Door Away From Heaven" by Dean Koontz Pg 8

to embrace change and take chances on what the future may bring. But if we embrace change, we can renew our culture and solve many of our problems at the same time. The old Native American saying, "We didn't inherit the earth; we've borrowed it from our grandchildren," is still profoundly true, yet we live borrowed to the max, as if there is an inexhaustible supply of credit, clean water, and breathable air. Kinsey Milhone, a private detective in Sue Grafton's novels, echoes the idea that we don't own all the things we use when she says "....life is only ours on loan."[8] In short, we need to get busy and start living like the intelligent species we are before our lives are over. After all, I'm sure we can find some fresher cheese out there in the "Maze"!

---

8 Kinsey Milhone in "K is for Killer" by Sue Grafton

Important Concepts of this Chapter

What appears ethical depends upon our worldview as we make decisions. Decisions based on the narrow perspective of I, me, now lead to huge collective bad results like our current social, environmental, and economic crises.

A new ethic must include a time, species, global perspective on all decisions. The resulting ethical basis will allow humans to persist in the long term as a biological species enjoying the abundance of earth.

Our spiritual lives become richer and deeper as we transition from behaviors associated with a narrow perspective to those that follow from an expansive perspective.

Humans are one species out of 1.8 billion named and known obvious species on earth. Earth is one planet of nine orbiting one sun out of billions in the Milky Way, which is one galaxy out of billions in the Universe.

The universe is made up of only 4% visible matter: the rest of the universe is made up of dark matter and dark energy that go straight through visible matter, including humans.

The bubble of unreality in which modern Americans live is only 0.000003% of Earth's history. We cannot make ethical decisions by ignoring all but one of the species on earth and by ignoring how things function in this ancient place nor by ignoring how our behaviors affect other species, societies, and ecologies around the globe.

* * *

The next chapter takes a closer look at the bubble of unreality in which we Americans live. Never before in history have people been subjected to such an aggressive, ubiquitous, and intrusive set of social signals designed to appeal to our basest instincts. With this system we have become slaves to consumption and debt. A key driver of the system is the central importance of economic growth. The first step to free ourselves from oppression is to become aware we are oppressed.

Chapter 2

Obligate Consumption

Years ago I read an unforgettable short story involving robots—the story was about what appeared to be a dream wedding. There were 12 limousines carrying all of the family members and friends, hundreds of robot servants, luscious spreads of every food and drink imaginable, the bride and bridegroom were dressed in elaborate attire, and the house to which they returned was like those being built today by the oil sheiks of the United Arab Emirates. This was a time when energy supply, resource depletion, and pollution problems had been solved, all of the work was done by robots, and in addition to the normal rules of robots there was an additional societal rule that nothing could be wasted. In this society, it became work to use, eat, or wear out the abundant goods produced by the robots. Toward the end of the story, it became apparent that the poorer the people, the more they were obligated to consume—the wealthiest people lived in what we would call dire poverty, consuming only the minimal amounts of food and consumer goods they wanted. One middle-income family had bought robots to work in the basement to wear out their Levis so they would be free to wear their old comfortable ones, essentially cheating the system in a logical way.

The situation of the poor in that story probably sounds much more appealing than it really would be—being force-fed consumer goods and services sounds like a horror story to me.

Wouldn't it be awful to be obligated to put on new clothing several times each day? I don't know about you, but I have some favorite clothes that I wear repeatedly because they are comfortable and bring back good memories. What about your college sweatshirt? Or maybe your running shoes? Or your favorite arm chair? What if you *had* to accept new ones every day?

In this chapter, I'm talking about consumption in its broadest terms. Consumption includes the act of eating or drinking, the act of using something up, as well as consumer expenditures that involve the purchase of goods and services. Ironically, another meaning of the word consumption applies to wasting diseases, a concept similar to what is happening to our cultural and environmental heritages as we engage in the excess consumption we've developed in the US.

As a first glimpse of the American social pressure to consume, think of the anxious articles in the business section of the newspaper as Christmas approaches: will American consumers save the economy by once again putting themselves in debt until the Fourth of July? What can you buy and give to your loved ones for Christmas that will show them how much you care for them? How many gifts must you buy out of social obligation? Predictably, American consumers are also obligated to overeat during the "joyous" Holiday Season and then buy diet products far into the spring. Possibly the science fiction story above is a testament to how excess consumption of goods can be horrible? No wonder some people bemoan the loss of the "real meaning of Christmas" as Christmas became an orgy of obligate overconsumption.

Since WWII, U.S. culture has evolved into a global oddity, a culture of obligate consumers whose members have developed levels of conspicuous consumption that is essentially a peacock's tail among human cultures. This hasn't occurred by accident—it is a product of free market capitalism and a public sold on the idea that increased consumption always improves the quality of life. Americans believe this because we live in an onion-like social structure of nested influences that drive us to ridiculously high levels of consumption.

The Onion, Layer One: Structural Consumption

At the core of the onion is a baseline of wasteful consumption that we cannot control without a great deal of effort. One way to get a sense of structural consumption is to pause at the curb or dumpster when you've taken your trash out for pick up. How much of what is in your trash container did you purposefully buy? If you are like most people, I suspect just about everything in there is a byproduct of living in the U.S. On average, each American produces about 4.5 lbs of trash per day, the world's highest level ever. As individuals we don't do this purposefully. Our trash production is a symptom of the cultural system in which we live.

Much of this excessive consumption occurs because the physical infrastructure surrounding us has developed in our free market system where conservation of resources has not been a consideration. Another way of saying this is that by living in our society, we have tons of collateral consumption—byproducts of how our buildings, communities, and social norms are organized. Toilets, paint, carpet, supermarkets, furniture, clothing, cars, medical care, restaurants, airplanes, schools, landscaping, soaps, and social expectations are all part of this excessive use of resources.

One example is water use and water waste associated with our homes. Building planners count on 150 gallons of water per person per day, 50 of which is used for toilets and showers. Think about this. Each time one of us has to urinate, 3-6 gallons of water is flushed down the toilet. Because of U.S. health laws, the water entering the toilet is clean enough to drink, a fact that most of our dogs know is true. Once we have mixed urine or feces and toilet paper into it, we have created a soup that must go through all kinds of water treatment before the water is again usable. Added to the slurry from our toilets is also the water used to do laundry, take showers, wash dishes, and other domestic uses. All of this together is sewerage, which, because of things going down toilets, is a serious health hazard. EPA-approved sewerage treatment plants are monumentally expensive to build and usually run near capacity.

In addition to our indoor misuse of water, we also habitually abuse the water that falls from the sky because of how our infrastructure is designed. Precipitation is a windfall for the plants and animals living in each spot. But we have turned it into an expensive problem called storm water that is so polluted that we wouldn't want it mixed with our source of domestic water. By replacing earth's natural coverings with asphalt, shallow-rooted turf grass, and other impervious surfaces, we have turned the water that would recharge the aquifers below us into streams of polluted water that must be removed from our communities. Then, in a poorly thought out I, me, now engineering feat of the past (another frozen accident), we hooked storm sewers to the sewerage system so now when it rains hard there are massive overflows of raw sewerage into our lakes and streams. Many U.S. cities are working to correct the overflows by building multimillion dollar systems to carry storm water separately from sewerage.

But you can't solve problems with the same level of thinking that created the problem. That is true is spades in this case. The separation of storm and sewerage systems now allows us to transport our artificially created storm water directly to our streams so we can pollute them directly. I suppose that's efficient, but certainly not productive. Thinking people have to look at the whole thing as a comedy of errors brought to us by past decisions that ignored water's value.

We originally did our toilet business in out houses, cold in the winter but effective for their purpose. Then we developed indoor plumbing—but washing the toilet was a nasty job, so the toilets were designed to wash themselves after every flush. Thus, six gallons or more per flush. Then federal regulations cut the six in half because of increasing water shortages and sewerage treatment system capacity. Now toilets are available that flush with 0.8 gallons of water. Actually, these toilets have two buttons on top—regular and super—the super uses 1.2 gallons. These new toilets are great. You know how the water swirls lazily around in a six-gallon-flush toilet? It gives you the opportunity to see all sides of anything you put into the toilet. Nice. When you flush a 0.8-gallon-per-flush toilet, you need to stand back because it sucks everything down immediately. Whoosh, it's gone. And you can set up whole cities with these kinds of toilets. The pipes will carry the lower water flow just fine. One more step can be made—composting toilets that use no water. I suspect most of us aren't ready for that yet, but it's worth keeping in mind.

There are also waterless urinals for men to use. These urinals are designed so that there is no smell because a layer of oil keeps the odor from coming up the drain. Think about how much water would be made available for humans and other species if we would just move en mass to decreasing the water waste associated with toilets! Think also about the reduced volume and cost of sewerage treatment each year. This is structural consumption because when you have to go, you have to go and the volume of water used isn't something you can control.

Another piece of water waste in the US is storm water, again a piece we cannot control by ourselves. Storm water runoff is a man-made problem that doesn't need to exist. As I mentioned above, it's a product of bad thinking (or no thinking) when the infrastructure of our cities evolved. Rain is a key resource for the land, even the land in cities covered by buildings. The goal of any construction should be to use every drop that falls on a property to take care of the needs of that spot. We can't control storm water runoff from our sidewalks and streets, but we can control it from our houses and yards. Our houses can be built with vegetative roofs that absorb water and reduce runoff—a bird or insect flying over the building would look down and see available habitat instead of toxic shingles made of petroleum products. The layer of soil and plants weighs more, but it also insulates the building, reduces the city's heat island effect, and protects the roof's water-proof layers from exposure to the sun so these layers have a life of 50+ years rather than the normal 10-12. Notice the increased life of the roof decreases the structural consumption associated with the necessity to replace traditional roofing materials every 12 years or so.

Another spot where we are locked into structural consumption that pollutes water, creates storm water, wastes clean water, and wastes land is our lawns planted with shallow-rooted turf grass. By now you may think I really have it in for turf grass. If so, you're right. In my mind, the extent and "health" of turf grass (deep green, weed-free, insect-free, and rodent-free) is an indicator of unsustainable conspicuous consumption, especially when it is grown in areas that don't normally support grasses of that type. Ground-level ozone is high in turf-grass-rich areas because the two-cycle engines of lawnmowers are some of the most polluting internal combustion engines we can possibly use. Frankly, if I had a small child, I wouldn't let them set foot on a "healthy" lawn because it is so loaded with toxins and the air above it is so polluted, it would be a hazard to the child's development. But we live in a country where our laws require that we have a lawn and we must keep it mowed.

There are other ways to deal with the land around our houses that help us out of the trap of structural consumption. One way is to build a rain garden[9], a depressed area of ground set up to collect water when it rains. A rain garden should be designed with native plants that live in a variety of water-availability conditions and also designed to hold water 10 days or less. Why 10 days? Because mosquito larvae hatch on the 10th day. Patricia Pennell of the West Michigan Environmental Action Council has popularized and promoted this concept widely.[10] Rain gardens trap pollutants and allow clean water to soak into the ground to recharge the aquifers. They also provide habitat for insect pollinators—by making space for good bugs, we have a lot fewer problems with the bad ones.

Structural consumption doesn't stop with our homes. Similar arguments can easily be made for the structural consumption built into the infrastructure that supports our communities. Currently, the only way electric companies make more money is to sell more electricity, most of which is generated

---

9 www.raingardens.org

10 www.wmeac.org

by huge coal-fired or nuclear power plants. Obviously it is in this powerful (pun intended) industry's interest to ensure we each use as much electricity at possible. Features like huge night safety lights that are wired without an off switch follow directly from the profit motive.

Have you ever tried to save money by using less water or electricity? If you have, I'm sure you noticed it's almost impossible to save much. This is due to the escalating price structure associated with our basic needs. The first water or electricity you use is very expensive. If you were to use 1,000 times as much, you would find the uses above normal household use are much cheaper per unit and almost free at the highest consumption levels at off-peak times of day. This is because our electric-generation and supply systems are set up to meet the maximum demands that might occur but can't be shut off or turned down easily when demand is lower. What other outcome would you expect from a social system set up to ensure maximum profit from maximum consumption rather than maximum profit from meeting consumer needs with the least impact on resources used and pollution produced?

The Onion, Layer Two: Marketing and Eating

Around this core of structural consumption is a layer of marketing. Individuals are barraged moment-by-moment by sophisticated marketing and advertising pitches designed to reach us as we go about our daily lives. We may not be aware of it, but there are thousands of highly paid professionals among us whose sole job is to convince us we cannot live without purchasing particular consumer items. They prey upon our insecurities, our biological urges, and on our desire for happiness.

One of the earliest TV commercials had to do with Listerine mouth wash—the message was that those around you knew if you had bad breath even if you weren't aware of it. This was a pioneering use of everyone's secret paranoia to sell a product. I remember those ads—after seeing it once even we children started blowing into our hands and sniffing to tell if our breath smelled bad. The ad was highly effective at the time—it created a need for a product when formerly no one had it and no one knew they needed it no matter how bad their breath was. Before Listerine, there was no social expectation of fresh breath; Shakespeare said "in some perfumes is there more delight/Than in the breath that from my mistress reeks." Also use of the product wasn't particularly healthy because it killed the normal oral bacteria that protect our gums and teeth from disease. This story has been replayed repeatedly in the years since with every product and cultural opportunity imaginable. Think about weddings, births, deaths, graduations, Christmas, Easter, Mother's Day, Father's Day, Valentine's Day, and so on. The list is endless.

Starting with the Listerine ad, global advertising turned into a massive business enterprise, a whole sector of the U.S. and world economy. One popular career specialty within business schools of universities is marketing and advertising. A whole culture of advertising has developed, much of which is a distraction, distortion, or deceit perpetrated on unwitting consumers. One poignant example is the plethora of weight loss products currently on the U.S. market. These products are advertised picturing formerly obese people who now are less overweight. The associated verbiage invariably uses the word "may"—"Our pill/drink/meal plan *may* cause you to lose X lbs in Y weeks. Try it free for a month and you will see..." The word "may" means "possibly"—the meaning of the advertising sentence is exactly the same if you substitute *may not* into it. Also nothing is life is really free. We will pay one way or the other. If that isn't the case, the business making the offer would go bankrupt and be "out of business" without the vulnerable to pay them for their useless product. I call the product useless because there is only one valid way to control our weight—eat less (preferably a balanced diet) and move more.

There is a close parallel between our personal money budget and our individual caloric budget. This is something I realized only after listening to Dave Ramsey, the Nashville get-out-of-debt radio host who is on more than 400 radio stations around the U.S. and also on FOX business TV. One day Dave Ramsey was talking about using a home equity mortgage to pay off credit card debt—he said this was like financing Saturday night's meal for 30 years—not a good idea. Building up financial debt is something that causes a constant hemorrhage of our buying power. Paying interest on money spent before it is earned reduces the amount we have to spend. Dave Ramsey makes it clear to his listeners that we have much more money and can become wealthy by living within our means, saving to pay cash for all purchases except our home, and being purposeful about every penny spent. One day it occurred to me that this is an exact parallel to weight gain, which is another symptom of overconsumption.

If we succumb to the advertising signals around us and overeat, it is just like overspending our body's energy budget. We build a debt in the form of unused energy that manifests itself as rolls of fat. We pay heavily for this debt with diabetes, osteoarthritis, and heart and kidney disease as well as by difficulty in moving our bodies with all that extra weight attached.

Just like credit card debt, there is no magic bullet to solve the problem other than paying back the debt. Losing weight is like paying for eating you did two or twenty years ago. If we apply Dave Ramsey's message to this situation, then people would eat carefully with health, nutritional, and esthetic purpose plus arrange their lives so adequate exercise is built into their daily routine. To do this isn't easy—a person must make choices that are atypical of mainstream American culture. One must also resist the barrage of advertising that leads people to choose the typical American diet and lifestyle, a path that has enticed 40% of Americans to be obese and the vast majority to be overweight. As other countries have adopted U.S. foods and cultural practices, a rising tide of obesity is spreading around the globe[11].

The Onion, Layer Three: Quarterly Earnings

A constant barrage of advertising marks our lives as Americans but the advertising is driven by the structure of our economy. The next layer outward of this onion-like structure of cultural influences is corporate competition. Once people cannot live without mouthwash or some other product, various corporate entities look to increased sales of that product to boost their quarterly earnings. Other corporations jump in to mimic success so there is a bewildering array of similar products we must choose from. There is a whole field of study associated with research on cues that draw consumers to buy—notice the research is not on quality of the product but rather on how to get us to buy it.

---

11 Cathy Newman and Karen Kasmauski, "Why are we so fat?," National Geographic, August 2004.

Quarterly earnings are the heartbeat of the U.S. economy—they are vital to the CEO's annual reward package and tied to capital growth of corporations. Corporate capital growth is tied to measures like the stock market and to the "value" of pension funds and other investment instruments that buy stock. Corporate competition can be ugly—it's important to remember corporations are competing to make the most money, not to provide the best products nor to meet people's real needs. Thus, the product frequently doesn't increase the wellbeing of the person who buys it, but it always costs the consumer as much as the market will bear.

A good example of this corporate competition phenomenon is most of the shampoo and other hair products currently available on the U.S. market. First, one would think there would be many different shampoos on the market, but that's not really the case. Between her bachelor's degree in chemical engineering and her graduate work, my daughter Adrienne had an internship with a globally known producer of various household cleaners and products like shampoo. Adrienne's job was to run chemical analyses on competitors' products to determine which additive had been put into the "new, improved" versions that continuously hit the market to boost sales. With this technique, corporations can avoid the expense of researching and testing new compositions for their products—they only have to follow others. If an additive creates major problems, the corporation quietly omits the additive in the next batch without reimbursing or telling customers of the error. If problems are minor and the change increases the profit margin, then the product will be sold. I hope you see what's going on—all the shampoos on the market are made from the same basic ingredients. Don't trust me—check this for yourself. Start reading shampoo ingredient labels. Look for one that doesn't contain sodium lauryl sulfate (or a near relative), an ingredient known to cause hair thinning and loss.

If the shampoos on the market were good for the health of people's hair, there wouldn't be a big problem. But this isn't the case. The common formula for shampoo causes scalp itching, loss of hair, and limp, dry hair. I've even heard of dermatologists who recommend Mane 'n Tail horse shampoo to help people with pronounced hair and scalp problems. That's sad, isn't it? Who would ever have thought we would live in a culture where horse shampoo works much better than human shampoo in stopping hair loss and improving the health of our hair?

Is thinning hair and hair loss a problem for the shampoo makers? Not at all. Actually, they make more money if people have problems because it gives them the opportunity to sell hair conditioners, gel, mouse, and hundreds of other products designed to help people handle their thin, damaged hair. Notice we are driven to buy to solve problems created by things we've already bought. And the system is set up that way to drive ever-increasing consumption in service of corporate profit growth.

The Onion, Layer Four: Planned Obsolescence

At the outermost layers of the cultural onion are interacting economic forces of escalation, speculation, capital growth, and the industrial mindset. Our corporate culture has developed a system that revolves around ever-increasing sales of products and services that will appeal to consumers so they will buy. Alternatively, clever companies design their product so when an unwary customer buys the product, the company is assured of future income from the repair, maintenance, or built-in senescence of the product. Toasters used to be durable kitchen appliances lasting for the lifetime of their owners and then doing service for another family. Now toasters, lawnmowers, stoves, TVs, kitchen faucets, and other household items fall apart after a few years of use. Repair parts are not available and it is cheaper to replace the item than to repair it. If repair parts are available, they must be purchased as part of a large, expensive, modular item that must be entirely replaced—for example, products are designed so you must replace an entire motor-fan assembly of a furnace because some bearings wear out or you must replace an entire hot water heater because a heating element burns out. Clearly this type of design meets corporate need for profit rather than the consumer's need for repair of an expensive piece of equipment in their home.

The resistance of the auto industry to building electric and other fuel efficient cars is another example. Gas powered SUVs, cars, and trucks require constant maintenance—oil, spark plugs, air filters, oil filters, and lots of other dirty parts. Think of the greasy hands of your auto mechanic. Each of those maintenance parts creates a stream of revenue for car companies, a stream of resource depletion and pollution if you consider the full life-cycle costs of each item, and (here's the important part from a corporate standpoint) a constant stream of guaranteed profit for corporations involved. Electric cars and the electric/battery side of hybrid vehicles aren't like that. For example, the electric side of my daughter Adrienne's Honda hybrid is scheduled to get its first maintenance check at 150,000 miles! The gas side of her vehicle requires fewer oil changes too because it gets less use with the backup of the hybrid technology. Fully electric cars like the EV1 produced for a few years by GM were even better. Maintenance on them consisted of rotating the tires filling the windshield wiper fluid.

No wonder GM withdrew the EV1 from the market shortly after it was introduced! One way or the other, the more we must buy, the shorter the life of each product, and the more expensive the repairs, the more market there is for sales. To me that makes these products "cheap junk."

Life in the Onion

When I was a child, there was a stigma on products from China because we considered them to be "cheap junk." Now I've transferred that stigma to nearly all the products available in our massive stores. Those who succumb to advertising and to the U.S. cultural mindset of consumption walk around in a wonderland of floor-to-ceiling cheap junk—much now made in China—guaranteed to fill their houses and closets to the brim and to outfit our homes with appliances certain to cost a fortune in energy and water use. When the space runs out, one can always buy cheap plastic storage bins of various kinds and complex closet organizers to store more junk in small spaces. And when that no longer works, storage units are available at modest prices (as advertised on television!) or one can always buy a larger house. Popular magazine articles tell us how to organize our stuff, and if that doesn't work we can hire professional organizers. We even go into debt to acquire as much cheap junk as we can as long as the price is right.

The key criterion we Americans tend to use as we make purchase decisions is price. The wealthiest Americans will buy any piece of garish junk if the price is high enough and the rest of us will buy it if the price is low enough. Our stores are swamped with racks of ugly clothes made in other countries, piles of bedding with clashing colors and loud prints, and rooms full of artificial flowers and plastic frogs—all at the expense of earth's health. There are more than 6.4 billion people alive on earth now—that's a lot of people, but more important yet is our ecological footprint, essentially our collective weight on the world. These piles of not-so-good consumer goods we feel compelled to buy, even when our buying puts us into debt, have global repercussions because our collective footprint is too massive for earth to sustain.

The piles of "goods" in our stores are being brought from around the world to the detriment of local ecosystems, both at the point of origin and here in the U.S. where they end up as trash, much of it toxic. Rainforest woods abound among our choices and are not even labeled as to the origin of the product. Cashmere sweaters come to us compliments of the great Asian steppes whose grasslands are degrading into desert from overgrazing of ever increasing herds of sheep. Gold is available from cyanide leaching processes that leave whole watersheds polluted and diamonds are mined from African countries while the local impoverished workers are subjected to full body cavity searches as they enter or leave the mines.

There is a wonderful book by Raj Patel called *Stuffed and Starved*[12] in which Patel provides detailed information about our current global food system. Living in our onion, we don't realize it but much of the food present in our grocery stores is exported from countries where some of their own people, including the farmers who grow the food we eat, are starving. For example, bananas are a common food most Americans enjoy frequently. What we don't know is United Fruit Company (founded in 1899

---

12 Patel, Raj. 2007. "Stuffed and Starved: The Hidden Battle for the World's Food System". Melville House, Brooklyn, NY

and now rebranded as 'Chiquita Brands') has been manipulating Caribbean, Central, and South American governments for over 100 years to control land, taxes, and labor to produce the banana supply for us. In the process, impoverished people are left without even the land to produce their own food and they are not allowed to protest the situation. One example Patel cites is Chiquita's recent payment of a $25 million as part of a guilty plea in its funding of paramilitary death squads in Columbia. Is there any wonder we currently have a flood of poor immigrants from the Caribbean, Central, and South America[13] (all considered to be "Mexicans" or "Cubans" by the American public)? This is where the food their countries produce is. Why not follow the food? And the jobs and money?

In almost all cases, the prices we pay do not reflect the social and environmental impact of the production of the "goods." For example, the Commodity Price Index (an economic measure consisting of a weighted average of the cost of energy, metal, and agricultural raw materials) has been steadily declining since the 1970s[14] even in the face of extinction rates that have reached ~27,000 species/year from habitat loss in tropical rainforests alone[15]. The CPI can decline in these times because with our current social, economic, and environmental policies corporations can externalize costs (avoid counting them as expenses) that damage earth's life support systems as well as the

---

13 Gorney, Cynthia. 2008. "Mexico's Other Border". National Geographic 213(2):60-79.

14 U.S. Bureau of the Census, Historical Statistics for the United States. 24 August, 2006.

15 http://www.pbs.org/wgbh/evolution/library/03/2/l_032_04.html, Retrieved March 17, 2008.

foundations of local cultures around the world. The profits thus gained from overconsumption and uninformed consumption contributes to the growing wealth of the privileged.

Up until recently, most Americans have been the privileged of the world. Our status as the most powerful and rich nation on the earth has insulated us from the negative social, environmental, and economic impacts of the exploitive practices that underpin the onion's existence. But now, with the global economic meltdown occurring in 2008 and 2009, most of us are starting to realize how we have been used as an unlimited source of profit for others. Our labor is now considered too expensive for many corporations but they still want us to buy their products with abandon. Fortunately, a majority of Americans are starting to understand how fragile our way of life is. They are also starting to resent excessive pay for corporate leaders and to recognize that their own prosperity has declined even while the media and government entities tout the strength of the American economy and ignore global threats like climate change.

A case in point is Gross Domestic Product averaged across the population. Ever increasing GDP per capita shows escalating wealth in the U.S. and many other developed and developing countries of the world at a time when nearly every global resource upon which our economy depends is declining. It's important to note that GDP per capita is deceptive because it is an average. Averaging 10 and zero computes to the same average as 6 and 4 or 5 and 5. Hidden in the ever increasing GDP figures is the fact that wealth is becoming more and more concentrated in the hands of a few.

For example, in the U.S. the bottom 20% has no or negative net worth while the top 20% own 96% of U.S. wealth. If you break down the top 20%, the concentration of wealth in the hands of very few becomes more apparent yet. The top 10% own 71% of the wealth, the top 5% own 56% of the wealth, and the top 1% own 33% of the nation's wealth. This isn't a "Bell-curve" of wealth, it's an "L-curve"[16].

To visualize how top heavy things have gotten, imagine you are attending a football game. The people of the U.S. are all attending this game and are seated in bleachers only on the south side of the field. Unlike the usual seating arrangement that has the wealthiest at the 50-yard line, they are organized in the bleachers by their income each year, the poorest at the zero yard line and the wealthiest at the 100 yard line. Yes, this is the wrong way to count yard lines in football, but pretend for a minute we can count them this way. Average income/year is delivered at the yard line in inches of $100 bills[17]. The 2007 data in Figure 1 below are from the US Census Bureau and include 95% of Americans. People on the 50-yard line are getting a little over a 2-inch-high stack of $100 bills each year and those earning more than 94.9% of their fellow Americans are earning a stack a little over 7 inches high. We all know some of us earn more than others so Figure 1 isn't any big surprise.

---

16 www.lcurve.org

17 Data taken from "Income, Poverty, and Health Insurance Coverage in the United States: 2007". Current population Reports: Consumer Income. U.S. Census Bureau. Issued August 2008. http://www.census.gov/prod/2008pubs/p60-235.pdf

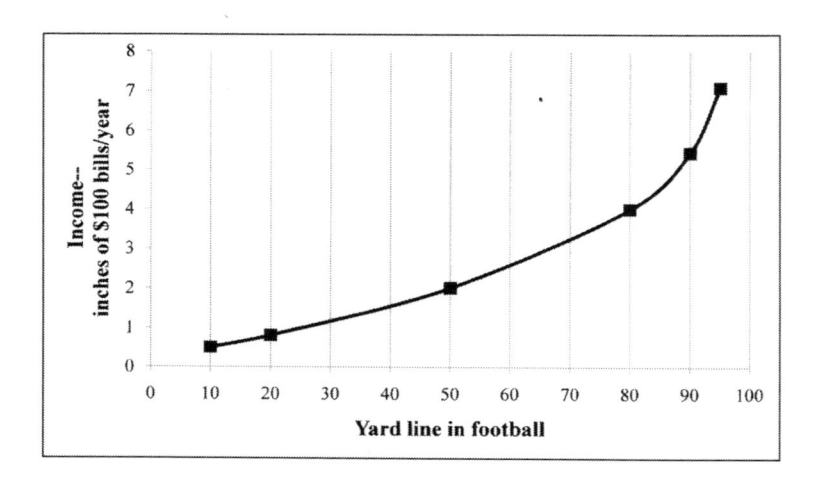

Figure 1. Comparison of incomes of the lower 95% of Americans. Data are presented as height in inches of $100 bills/year and income at the 50-yard-line represents the average income of Americans.

Now let's take a closer look at Americans seated above the 95-yard-line. Their yearly incomes are so much higher than the rest of us that they can't be graphed on the same scale as our incomes. In Figure 2 below, I've graphed the yearly income of top earners with a scale appropriate to their income levels[18]. I use the word "earners" loosely here—actually these people don't earn their income in the usual sense, their wealth earns their income for them. Figure 2 is plotted in inches of $100 bills/year to make the two figures comparable, but 1,900,800 inches isn't a number that makes much sense to most people. It actually represents about 30 miles. Can you imagine try-

---

18 Data taken from www.lcurve.org and pertains to 2007.

ing to spend a stack of $100 bills 30 miles high? The inches of wealth for the richest is growing at a rapid rate. According to a comparison of Forbes listings in March 2006 and March 2007, there was a 19% increase in U.S. billionaires that year and their combined total wealth was $3.5 trillion, an increase of 35% over the previous year[19].

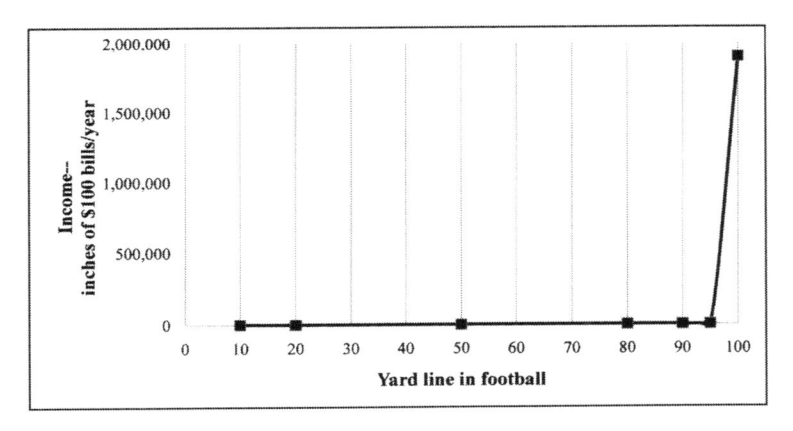

Figure 2. Comparison of yearly income of the top 5% with that of the bottom 95% of Americans. Data are presented as height in inches of $100 bills/year and income at the 50-yard-line represents the average income of Americans.

This is an amazingly top-heavy system and is repeated around the globe in nation after nation whose economies have been shaped purposefully to export local resources to meet Americans' ever increasing demands. In the meantime, savings as a percentage of disposable income in the US has declined

---

19 www.lcurve.org

from ~10% in 1980 to 0.4% in 2007[20]. Bankruptcy filings in the U.S. have increased from ~250,000 in 1980 to well over 1,500,000 in 2004[21]—these figures are worse now because of the sub-prime mortgage debacle and the subsequent global economic meltdown. Our collective government debt is also skyrocketing—at the moment of this writing in March 2008, it is $9,413,201,598,610.68.[22] That translates to more than $31,000 per U.S. citizen—our babies are born $31,000 in debt before they take their first breath! Talk about a culture of consumption that has evolved to a point of implosion!

We pay for our excess consumption of food, water, services, resources, and material goods not only with debt but also with our health and wellbeing. Our collective physical and mental health has been declining—our children bear the brunt of the problems now and will be paying for our errors throughout their lifetimes if we don't change our ways. Overconsumption is most visible in the food we customarily eat.

According to *National Geographic*[23], the average American ate 278 more pounds of food per year in 2000 than in 1970. Serving size for a Hershey's bar has gone from 2 oz (297 calories) in 1900 to 7 oz (1,000 calories) in 2004; McDonald's fries from 2.4 oz (210 calories) in 1955 to 7 oz (610 calories) in 2004; and Coke has gone from 6.5 fluid oz in 1916 to 16 fluid oz in 2004.

---

20 Time Magazine, March 24, 2008, "10 Ideas the are Changing the World," Source of Data: Bureau of Economic Analysis.

21 American Bankruptcy Institute

22 http://www.brillig.com/debt_clock/ Retrieved March 17, 2008, 08.07.06 PM GST.

23 Cathy Newman and Karen Kasmauski, "Why are we so fat?," National Geographic, August 2004

More than 30% of the U.S. population is obese and this trend is spreading around the world. Obesity in U.S. children and adolescents has almost tripled since 1980 and most schools have cut back on physical education because of budget constraints. The health impacts of food overconsumption coupled with lack of exercise include dramatically increased incidence of liver disease, colon cancer, osteoarthritis, stroke, type II diabetes, and heart disease. This is no way to treat ourselves and no way to raise children.

We don't fare much better with our mental health. The Center for a New American Dream,[24] which has studied the attitudes of U.S. citizens, reports that well over half of Americans recognize their consumption patterns cause them to work more, have heavier debt loads, lose time with their families, and cause serious environmental harm. Even 15 years ago on our escalating consumption curve, the data indicated a severe loss of well being coupled with increasing consumption. If we plot money spent against fulfillment, at the lowest income levels earning and spending money dramatically increases fulfillment because it allows us to survive. Adding more earning and spending to the situation even provides comfort on top of survival thus clearly increasing fulfillment. Increasing earning and spending to a higher level yet adds luxuries on top of comfort but adding luxuries on top of luxuries does little to increase fulfillment.

As a matter of fact, adding too many luxuries becomes stressful because one generation's luxuries become the next generation's necessities. For example, large fries and a quart of softdrink look like big servings to parents but they look *normal* to our children. As another example, my parents' occasional treat

---

24 http://www.newdream.org/

of tuna fish became my generation's staple seafood. Meanwhile my occasional childhood treat of shrimp has become an everyday food choice for Americans. Bare feet in childhood gave way to simple tennis shoes, and those have been replaced by high-tech, imported, athletic shoes expensive enough to break the family budget. On average we spend more than nine hours per week commuting to work; when you take into account all of the time earning the money to pay for the costs associated with our vehicles, we travel about 2.5 mph, about the speed of a horse; in 1996, over 27% of Americans earning over $100,000/year felt they couldn't buy everything they needed; and we spend twice as much on children's shoes as we do on their books. Through these and multiple other "improvements" in our lifestyles, Americans haven't experienced any increase in happiness. As a matter of fact, by any measure we wish to choose—incidents of road rage or school shootings might be examples—Americans are less happy now than they were before this craze of consumption became the norm. I don't know about you, but I would feel anything but happiness if my child stomped through a mud puddle with fancy athletic shoes whose cost added to the interest I paid each month on a credit card balance.

    This treadmill of work-consume-work-consume hit home with me while I was Dean of Science and Natural Resources at Lake Superior State University in Sault Ste Marie, Michigan. This small university is located in a beautiful setting on a hill overlooking the international bridge over the Saint Mary's River that connects the U.S. to Ontario, Canada. In my career, being Dean was my first job where I was expected to show up to work dressed in business attire. At first, it was fun to shop for attractive business suits with coordinated purse and shoes, but after a while the story got old. It dawned on me one day that I went to work to earn money then spent my non-work time shopping for clothes to wear to work to earn more money. About that time, I opted out of the "clothes game" and shortly changed jobs to one where there was no "dress down day." Every time I attend a function where people are dressed up, it always reminds me of how silly the European royalty looked in the 1700s with their powdered wigs and outlandish costumes. Just because we all dress with a particular style doesn't make us look any less silly, in my opinion.

Fortunately, we don't live in a robot society where we have no choice except to consume. We can step off the consumption "party train" and step into a purposeful life that meets our material, social, and spiritual desires and take care of our health and the earth at the same time. As we make this kind of choice, we gain all sorts of freedom—the good news is we even get to choose the type of freedom we wish to pursue.

Those who use a time, species, global perspective to make decisions about their lives are the primary beneficiaries of their actions. With a broader worldview and a deeper spiritual connection to our place, it is easy to decide what's important and what is not. By de-cluttering our lives and moving from obligate consumers to purposeful consumers, we save money, live a healthier lifestyle, and avoid debt. We can live in a more comfortable house, increase our mobility, decrease our exposure to toxins, and avoid foods that do a poor job of nourishing our bodies. This sounds like a mythical promise, but it's not. The waste, rush, and rampant consumerism we've been steeped in are not foundational to our well-being. A healthy earth, peace, and sustainability *are* foundational to our well being, however.

Clearly we cannot achieve those ends with an I, me, now ethical framework.

At least for the time being, we Americans have sufficient excess wealth to give ourselves the freedom to carve our own mini-culture out of the options available. We can choose to live our lives as we wish and to opt out of the treadmill of consumption that has become our default obligation as Americans. Part of my purpose in writing this book is just exactly that—to hand you the basic information you need in order to make yourself truly free in the context of the natural abundance of our home here on earth.

Important Concepts of this Chapter

Since WWII, Americans have become a culture of obligate exorbitant consumers surrounded by a constant barrage of sophisticated advertising that preys on our insecurities and biological instincts.

Overspending any budget, financial or caloric, results in a form of slavery—either to debt payments or to the health impacts of weight gain.

Corporations exist to make money, not to provide healthy, useful, economical products for people to buy.

Financial markets and capital escalation drive our dependence on economic systems that must always grow to be healthy. This means we are trapped into buying through schemes of planned senescence, behavioral traps, and social mores.

Cheap products don't include social and environmental costs in their price.

Obligate consumption doesn't make us happier, healthier, or wealthier.

* * *

The next section of this book is about rethinking the core issues of our existence here on earth. First (Chapter 3) is an invitation for each person to mentally step back into being a welcome part of the rest of nature. As we mentally step back into nature, we must recognize the ubiquitous impacts of evolution that affect each of us all day every day. From the microbes around us to the course of disease in an individual to the trends in our culture, our lives are constantly being influenced and shaped by evolution occurring now. The second core issue (Chapter 4) is that humans are not intrinsically bad or destructive. Like all other species we modify our environment to make ourselves comfortable and to care for our children. But if we were behaving as a part of nature, we would design our activities so they don't undermine our earth's life support systems and our own well being. The third core issue (Chapter 5) is a discussion of key operational rules of nature that have developed especially for earth's inhabitants as an adaptation to how this place works. Because these rules have developed through the rigors of natural selection, using them as design principles gives us the opportunity to tap into the wisdom of the eons of earth's history.

Section II

Rethinking Core Issues

Chapter 3

People Are a Part of Nature

Although my father worked in a factory when I was older, he was a farmer and my mother was a farmer's daughter. The old time successful farmers lived their lives in tune with nature's cycles. Their sensitivity to these cycles bore fruit in terms of the productivity of the farm. So for me as a child the similarities between the species around our family and us was a given. It was only after I left the countryside of Michigan to pursue an education that I came in contact with the deep separation from nature most people feel. To gain a time, species, global perspective and to free ourselves from the consumption onion, it's vital to rethink that separation.

Mice

Mice are great teachers of the important issues in life. Take the field mouse that moved into my house last Thanksgiving. I'm from a big family and so I had a house full of company. The weather had been bad with a massive snow storm hitting the Midwest the day before Thanksgiving making things a bit frazzled for everyone. Thanksgiving night, after a big day of cooking, eating, and hosting family, my sister Mary Jo, her husband Kent, and I were relaxing while watching a movie.

Mary Jo and Kent are from Missouri and they live an exceptionally rural lifestyle so the sight of a mouse running along the wall in my house didn't raise any alarm with them. But when

I spotted the mouse, I jumped up, ran around trying to find the hole it used to enter and then set out some baited traps and some poison grain thinking that would solve the mouse problem. My visitors thought this was a lot of fuss and flurry. They had noticed the mouse earlier, thought my house was infested, but didn't find that much of a crisis. Obviously I did even though I am a deeply environmental person, not a classic tree-hugger, just deeply environmental.

That night I was sleeping on the couch to make room for my guests and was reading before going to sleep. Like the Christmas story, not a creature was stirring, not even a mouse, at least for a while. Then I noticed the mouse running across the middle of the floor. No skulking along walls for this mouse. Tail up, with a running stance like a stallion at full speed, this mouse was having a great time checking out the entire house. I lay there watching and could feel this mouse's exuberance…no snow in this wonderful new habitat it had found, plenty of food even though a mouse had to be careful to avoid the toxic stuff behind the couch and careful to avoid contraptions that didn't look safe. To my horror and fascination, the mouse finally jumped up onto the couch with me, hopped up to the back of the couch and then shimmied up the vertical blinds, ran back and forth along the top of the blinds, and back down to the couch. It raced up and down on top of the covers over me and finally noticed I was there. It came right up to my face and, standing on top of the covers over my chest, looked me right in the eye.

This was the most beautiful little mouse I had ever seen in my life. It wasn't a dull grey house mouse with their nasty smell. It was a dark brown mouse with a long tail and big, shiny brown eyes. The bottom half of its tail was pure white, a cue that it was some type of deer mouse that normally lives in fields. He (or she?) wasn't a bit intimidated by me. After our meeting, I was feeling a bit guilty about the poison and traps, but I'm prejudiced enough about mice to be sure you can't have them living in your house without accepting a huge mess so I left the killing stuff in place thinking guiltily that he would eat it and disappear to die. In the meantime, I left him to continue his explorations, which included tearing up and down my body repeatedly. Finally he found something else to explore and I went to sleep.

The next day my company left and my first order of business was to make sure that mouse was out of my house. I cleaned like a maniac, flipped the couch upside down, and hunted thoroughly. No mouse anywhere and no evidence that it had touched any of my mouse killing equipment. I thought it had probably figured out that it wasn't welcome in my house and had exited the same way it entered. But that's not what the mouse was thinking.

I have many animals in my home; most of them are creatures displaced by human activity and in need of a place to live. In particular, I was on the couch because my second bedroom belongs to a pair of green iguanas that sleep on a carpeted shelf near the ceiling and spend their days sunning under a full-spectrum light bulb. As is my habit, late in the afternoon I went into the lizard room to turn off the artificial sunlight and to straighten up a bit. To my amazement that day, the two four-foot-long lizards were hiding up on their shelf and the mouse was basking under the light and eating the lizard food. He looked so happy. Not only had he avoided the snow and cold outside, he had also found a place with abundant food where it was constantly sunny. And all he had to do was to chase off a couple of frightened lizards to have it all to himself. Clever and cute or not, I immediately declared war.

I stuffed towels under the doors and tried to catch and kill him, an amazingly futile exercise. We were both so traumatized by that war that I decided to change tactics. Finally I built a barricade close around the sunlight corner, trapped the little fellow in a cottage cheese container, and hauled him outside. When I turned him loose, he burrowed quickly into the wet, soggy, snow-covered grass with a disappointed backward look at the house.

Now some would say I "anthropomorphized" this mouse, which is just a fancy way of saying I have projected human thoughts and feelings into him that mice do not have because mice are vastly inferior to humans. In this worldview, humans were created separately from the rest of nature and have total power over all of it. Taking a quite different view, Frans deWaal has proposed the term "anthropodenial"[25] to describe our denial of our close shared characteristics with creatures like mice. We eat and choose similar foods, drop turds (fortunately his were tiny and dry!), drink fresh water, mate, have babies like ourselves, and move into a "home" that is the best we can do for ourselves.

Yes, there is a difference of specialization between mice and humans, but the same essential features are present in both because of our shared ancestry through the process of evolution. Both humans and mice are constantly and continuously being shaped by the forces of evolution, as are all other living things on earth. Because evolution is such a sore, misunderstood spot with so many people, I'm going to spend some time on the practical and philosophical issues related to evolution because you cannot have a true time, species, global perspective without understanding evolution's inevitable impacts.

---

25 "The Ape and the Sushi Master", Basic Books, 2001

The Evolution War

Just the mention of evolution is a fighting word for many Americans because evolution has come to symbolize the loss of God from our purpose, our lives, and our civilization. In the minds of many, this loss has corresponded to the tremendous increase in money, power, and influence of the scientific endeavor. It seems now that science is sticking its ugly, know-it-all nose into everything. Science's "objective data" appear laden with covert philosophical messages that threaten the core of traditional religious beliefs and the very meaning of humans. And the loss of these beliefs threatens the moral capital upon which our civilization is built.

Similarly, just the words biblical creation, or "intelligent design" as creationism is now being championed[26], brings out the distain in many scientists, especially biologists. Many biologists feel like Repairman Jack, a character in F. Paul Wilson's book, *Bloodline*: "The good old creationists, sabotaging knowledge wherever it rears its ugly head." Biologists value ecosystems, mice, and all the species on earth as our heritage that has developed through 4.56 billion years of inorganic and organic evolution. Those who know the history of earth have seen no evidence of a quick, magical creation so they recognize that nothing but vast spans of time could recreate the wonders of earth's biodiversity if it is lost. Biologists also understand how humans rely on earth's biological capital to underpin our lives, economy, and culture.

You wouldn't think the two camps would agree on anything, would you? But most members of both camps agree that humans

---

26 Brumfiel, G. 2005. "Who has designs on your students' minds?" Nature. 28 April 2005. 434:1062-1065.

are above nature or not part of it at all. Christian doctrine is almost universal in its proclamations that earth was created for the use of humans and that humans are the pinnacle of creation. Even among Christians who accept evolution and among many biologists, the science is twisted to become a progressive mechanism that leads inexorably toward humans; frequently this concept is coupled with the assumption that the story that has unfolded on earth for the past 4.56 billion years has ended because humans are finally here. Biologists, and ecologists in particular, adamantly deny human-altered landscapes, human technologies, and human material goods as being natural even though around the globe it's possible to tell immediately which areas are inhabited by humans just as you can tell the stages of ecological succession, the predictable sequence of biological communities that replace one another as the group of organisms living in a local spot evolves into a stable climax ecosystem. I would agree with ecologists that human-dominated areas haven't reached an ecological climax state—that is, human-dominated communities haven't evolved to a steady state condition where nutrients are cycled through the ecosystem and energy flows through trophic (feeding) levels, which is what occurs in non-human biological communities that have reached a steady state condition. At the same time, I would point out to the ecologists that the non-climax stages of non-human ecological succession are similar to the human ecosystem in that they haven't yet reached a stable equilibrium either. Regardless of the ecological details, most people consider humans and their activities to be separate from nature.

People on both sides of the evolution argument are highly invested in preserving the things they value and the longer the argument continues, the muddier the waters get. Some religious groups are putting forward religious beliefs as though they are science and many people, including at least a portion of scientists and teachers, are putting forward philosophical materialism as though it were science. Some have adopted science as though it is a religion and others have adopted religion as though it is a science. The rhetoric is spiraling downward and the damage to humans and the earth is spiraling upward. Maybe the negative tension is due to false mindsets rather than to anything of substance?

John Haught[27] and other theologians, philosophers, and scientists who spend their time thinking about the interface between science and religion have pretty well agreed that this war is unwarranted. The war between science and religion or, more personally, the war between evolution and creation, exists because these things are seen as alternatives to one another. In this war between atheism and fundamentalism, either (a) God created us in his own likeness and made us stewards with dominion over the earth, setting us aside as special creatures who alone on earth have a moral imperative, or (b) this universe is a lonely place with no God and it developed through materialistic, cosmological and Darwinian mechanisms that don't know or care whether humans exist or not. If we accept this binary mindset, we have a difficult choice. Either we can nurture and value our moral/spiritual lives or we can forget a higher power and spend our time seeking naturalistic explanations for our observations. In this either/or situation, we lose either our moral/spiritual foundation or our ability to use science to

---

[27] John F. Haught. "Science & Religion: From Conflict to Conversation", 1995, Paulist Press

understand this world. But I'm convinced all people on earth will lose if we accept this false dichotomy.

Each of us, from the most materialistic scientist to the most fundamentalist religious person, has both scientific thinking and spiritual thinking as integral parts of our psyches.[28] To deny one or the other of these is to deny a part of ourselves. In addition, we all lose because, if science is not informed by moral/spiritual thinking, the scientific enterprise can be turned to the most rapid and effective destruction of our social and natural capital imaginable. Because our science now is not well informed by moral/spiritual values, much of this century has fostered a scientific establishment that developed the atomic bomb and the wherewithal to deliver and detonate it plus a scientific establishment that has developed methods of food production that victimize the soil, our water supplies, our food animals, and the humans who eat cheap, abundant, mass-produced food that barely nourishes the body, let alone the spirit. When our spiritual lives are not informed by science, our society propagates economic development strategies, medical techniques, pest control, and agricultural methods that ignore the inevitability of natural selection in our biological heritage. In both cases, our moral/spiritual imperative with respect to one another and to the earth is not met *even when we think we're doing the right thing.* We harm ourselves, each other, the generations of the future, and our fellow inhabitants of earth.

This isn't how things have to be. John Haught says we actually have four choices in how we relate our moral/spiritual selves to our scientific selves.

---

28 John Polkinghorne, "Belief in God in an Age of Science", 1998, Yale University Press

Philosophical Options

The first choice is Conflict, the either/or philosophical stance demonstrated above and popularized by Dr. Dino[29], the late Pat Robertson, Richard Dawkins, and the media. This option assumes the only source of moral/spiritual thinking is from established religions and leaves us with no choice but to reject the opposition and pay the consequences. When our children are taught conflict, they must subvert either their science or their religion if they are to develop a unified worldview as they mature. This leads to people who are either impoverished spiritually or people who think they are acting in moral ways as they make I, me, now decisions from a narrow, human-centric framework. Fortunately, conflict isn't the only option we have.

A second option was put forward by the late Stephen J. Gould, who relegated science to one domain of our thinking and religion to another[30], a philosophical position Haught calls Contrast. In this metaphysical view, we recognize our spiritual/moral meaning and at the same time recognize our physical heritage as it is being discovered through the sciences. In Gould's perspective, there are "things of the spirit" and "things of the natural world." God isn't tampering with the natural world and we cannot look to the natural world to define our moral responsibilities. Gould felt the humanities, organized religions, and the arts are liberated with this view and left responsible to develop an understanding of God, human purpose, and spiritual responsibility. The sciences are left with the responsibility to honor the limits of the data they collect as devoid of moral

---

29 http://www.talkorigins.org/faqs/hovind/

30 Stephen J. Gould, "Rocks of Ages: Science and Religion in the Fullness of Life", 1999 Ballantine

meaning and spiritual directives. Thus, philosophical materialism is not supported by the mechanistic explanations of science, and we can live our spiritual lives based on the tenants of deep religious belief. Separate domains of human experience and meaning are the hallmark of Contrast.

For some, especially Christians, the schizophrenic thinking of Contrast is inadequate. These people recognize the purposelessness (in the narrow sense of human purpose) of natural selection has implications for our spiritual beliefs and also the assumptions under which science are conducted have an impact on our thinking as we develop theories used to organize scientific data. For example, a pillar of Western science is uniformitarianism, the assumption that the processes we see occurring around us are the same all over the world and were the same in the distant past. This assumption is based on monotheism, the idea that there is one god who is ubiquitous. If modern science had had its origins in the Orient where religious beliefs include a series of gods in local forest groves, then uniformitarianism would not have been assumed. Assumptions like uniformitarianism underlie all scientific theories including atomic theory, evolution, and relativity. Of course, scientific theories show up in our schools and impact how people think, forming the assumptions that underlie their beliefs. With the philosophical stance Haught calls Contact, interaction and discussion and attempts at reconciliation occur in the interfaces where science and religion come in contact. The point of the discussion is not either/or, who wins, or conflict. Instead the point is to cooperate to seek deeper understanding and harmony, or at least lack of conflict, between these important aspects of the human endeavor. In reconciling evolution and creation, a person with a Contrast philosophy might believe in theistic evolution—that is, evolution is God's creation method. Or they might agree that there are evolutionary reasons for humans to develop moral behavior and belief in gods even while they continue to practice their religious faith. Contact is a stepping stone to the worldview I espouse, that of Confirmation.

Confirmation starts with the premise that there can be no real contradiction between valid spiritual beliefs and valid results discovered by science. When contradictions occur, there is an error somewhere, but we must be careful to avoid using only experiments and data or only theology and religious interpretations to determine what is correct. Ideally, the truth we discover through reason, logic, and science confirms the truth we discover through religious studies, meditation, and inspiration. Our spiritual lives can inform our scientific minds of directions to pursue and methods that are morally acceptable to use in the pursuit and our scientific discoveries can open new avenues for our spiritual understanding. In this metaphysical view, evolution is not a threat to our religious beliefs but is instead an observation that helps us to deeply understand this world. Thus, we become responsible to pursue moral solutions to problems in a way that synergizes with the natural world rather than crushes the raw material upon which it functions. We are also responsible to pursue social and moral goals that embody our role as the one species on earth that is looking inward and outward to understand our purpose and to understand ourselves, our spiritual obligations, and our home on earth. In the Confirmation worldview, there is no conflict between informed religious views of creation and the details of evolution. Thus, the harmony and logic of the Confirmation stance leads to well-being for humans and for mice too.

If I've offended your religious views with any of the above, I apologize. But this isn't a book on religion; it's a book about how we can perceive the world in a way that leads to a sustainable harmony between humans and the rest of nature. To get to that harmony requires a time, species, global perspective and that perspective cannot be attained without understanding evolution as the key underlying mechanism that impacts and underpins our well-being.

Evolution—A Simple View

Whether we are aware of it or not, evolution is working on us and around us all the time. Here I'm talking about evolution in the broad sense of the term, not just biological species-making, but also any change that is fed by new variations and sorted by selective forces. Think back to when you were a child and envision your neighborhood. If you've visited that neighborhood since then, I'm sure you've noticed that changes have occurred. For each change, someone or something has selected a particular option over others available. Some trees have grown, others are gone. Neighbors have gotten older, some have moved, some have died, and new children, weeds, and trees have started their lives. In all cases, from the mice to the humans to the weeds, some of the offspring lived and some died. Many more eggs, sperm, and embryos were formed than the fortunate offspring living there. Humans are a part of this—about one in five human pregnancies ends in spontaneous abortion so selection is working on the people in the neighborhood too. There's no reason to fight it, that's just how things work in this place.

Evolution really isn't anything mystical—it is an extremely simple process all about what works best in a local spot. Evolution that is occurring today in each place is functioning on the shoulders of 4.56 billion years of Earth's molecular and biological evolutionary history—that makes the evolution occurring today look much more complex than it actually is.

At a biological level, errors (mutations) can occur in the templates of life—the genes—or in the mechanisms that control the functioning of genes. If the error happens to be one that causes a detectible change in the whole organism, then it is screened by the process of natural selection. If the new variant works better *in that time and place* than the old version, then positive selection occurs and the new form spreads through that local population over time. On the other hand, the new variant may make no difference to how well the organism can survive and reproduce *in that time and place*—if that's the case, then selection is neutral toward the mutation so over time this type of mutation will accumulate in the population. The third possibility is the most prevalent, that is, the new variant doesn't work as well as the original in which case purifying selection will clean the mutation out of the population.

To illustrate this, imagine the elephants that were prevalent in the north when the ice ages began. Imagine also that some of the elephants had a mutation that made them bald, others had different mutations that increased the thickness and amount of hair they had, and still others had mutations that changed some of the basic components of hair but didn't change its insulating properties. Obviously purifying selection would take care of the bald elephants in this setting—they would freeze to death. Hairier elephants would do better than the normal elephants, so positive selection would spread their mutations as the climate got colder because they could survive and reproduce more. Finally, selection would be neutral toward the basic components of hair as long as the hair acted the same for insulation, so this change wouldn't affect the survival or reproduction of those elephants unless the new kind of hair made them too ugly to find a mate (negative selection in that case). In all of these cases, the elephants aren't better or worse elephants—they just are more or less adapted to that place and time. If we had been talking about elephants in Africa as the climate warmed there, the whole situation would have been reversed.

Evolution—A Modern View

Not very many years ago, scientists' understanding of evolution wasn't much deeper than the story of the elephants. But now, with the discovery of DNA (deoxyribonucleic acid), molecular tools to manipulate and sequence DNA, and systems to study the action of genes, it is possible to zoom in on DNA to observe the fingerprints of natural selection on the genes and the mechanisms that control gene action. This technology has shown us that mutations aren't mystical. Instead the majority are caused by regularly occurring energy changes in the molecules involved in the DNA code of the genes. In other words, mutations occur with associated probabilities that are, in a broad sense, predictable just like the probability of heads versus tails are predictable in a large series of coin flips. For DNA sequences, this means observed deviations from the predictable occurrence of mutations can be ascribed to the action of positive or purifying selection and agreements between the observed and predicted occurrence of mutations can be ascribed to neutral (or no) selection. This powerful observation allows us to reconstruct the evolutionary history of organisms from their DNA in addition to the traditional way using the fossil record.

These same principles can be used at all levels of biological relatedness from within species to between species and between larger groups. So at a broader level, comparing DNA sequences allows us to see the types of evolutionary forces that have been working every day, all day, throughout deep geologic history on the differences and similarities found. Because over time this system works almost like a ticking clock, it can even be used to determine an approximate time when groups split from their common ancestor. It has been used recently by Stephen Oppenheimer to track the migration of humans out of their ancestral home in Africa to the far reaches of the globe using mutations in the Y-chromosome, the sex chromosome unique to males. Oppenheimer's *Journey of Mankind* corresponds closely with and supports other DNA studies as well as the archeological record. This migration of humans was essentially a series of within-species splits of groups from their ancestral group. The molecular techniques are so sensitive that they are also being used to track modern-day evolution.

As the DNA of more individual humans is being sequenced, we are finding that many fingerprints of evolution in action can be tracked in humans. The December 21, 2007 cover of the prestigious journal *Science* has a t-shirt printed with the gene-sequence map of human chromosome 1 that symbolizes the conceptual advance in human genetics that occurred in 2007. We now realize that DNA differs much more from person-to-person than we had ever envisioned. If researchers keep at it, then eventually it may be possible in the future to do our genealogical trees through an analysis of our individual DNA! Currently, the degree of relatedness of groups is being determined by comparing the accumulation of shared and unique mutations in the DNA or in the proteins coded by DNA.

Evolution in Humans

In recent years, molecular data have fleshed out the information we have from the archeological record of the pre-written history of humans. For example, from fossil record data it was generally assumed that humans separated from their closest biological relative, the chimpanzees, about 12-15 million years ago. Much of that estimate probably stemmed from the wide gap we humans feel between us and the animal kingdom, so it was assumed it must have taken a really long time for such big differences to evolve. But early molecular clocks put the divergence date between chimps and humans between five and six million years ago. After a long scientific battle and after new significant fossil finds coupled with advances in molecular techniques, we now consider chimpanzee and human lineages to have separated only about five million years ago. When you consider human brains started enlarging about two million years ago and our modern brain wiring was mostly completed by 20,000 years ago, we must look at our nearest biological relatives, the chimpanzees, and consider them to be more important than just research subjects. We share about 97% of our genes with chimpanzees—the other 3% cause critical differences in our embryological development and gene control to create the obvious differences between us. The differences that exist between humans and chimpanzees are there because of the differences in the history of mutations and selective forces that have worked and continue to work on our respective gene pools.

Since our species shared a common ancestor, chimpanzees and humans have had very different success, especially in the past 100 years as the global human population has soared. Every continent on earth has a huge population of humans, most of who are thriving well enough to ensure the survival of the children they have. Chimpanzees, on the other hand, are restricted to ever smaller preserves in Africa—populations still surviving are there only because of human protection. In their evolutionary history, the mutations that occurred in chimpanzees and sorted by natural selection kept them adapted to relatively narrow habitats that occur only in Africa. Today, chimpanzees are not well adapted to the ravages of destruction of their habitat, collateral damage from human wars, poaching for bush meat, and murder for trophies to sell.

Evolution also occurs in humans in the sense that disease organisms can evolve within our bodies. An excellent example is HIV, the virus that causes AIDS. Once HIV enters the human body, it reproduces itself repeatedly by attacking specific white blood cells associated with the immune system of our bodies. Controlling HIV infection can only be done by a whole battery of drugs taken simultaneously because the virus is continually mutating each time it reproduces. The new mutations give the virus immunity to the action of one drug or another so by giving a whole series of drugs simultaneously, the virus and its tremendous mutational ability are kept in check for years. But all it takes is one mutant that can reproduce in the presence of all the drugs being used and the person will develop AIDS (immune system collapse) and will eventually die. Not a pretty picture, but evolution doesn't function as we wish, it is just an outcome of how things are here on earth.

The bottom line here is the creative system of evolution that has given rise to all this abundant life on earth isn't a process of the past. It's going on now and impacting our lives every day. And the fingerprints of our ancestors and the process of evolution are all over our genes and in the sequence of codes in our DNA. We're essentially permeated inside and out with the common ancestor impacts of evolution. For those with strong religious beliefs, there is no reason to cut a higher power out of this elegantly simple yet mindboggling complex system. Each of us is indeed unique and special, just like each individual of every other species. Instead of denying this simple yet elegant system, it can be used as a window into learning what a higher power is like. Think about the spiritual implications of a higher power that holds this universe in existence, set it in motion and endowed it with the ability to constantly, creatively renew itself through a process as simple and elegant as evolution.

Species Affect Each Other's Evolution

There's no way around it. Humans are firmly a part of nature and subject to the same forces of evolution that impact other species. The strange thing is that once life got started on planet earth, the main selective force that drove evolutionary change was species living in, on, and near one another, just as illustrated above with the fate of chimpanzees in Africa. Don't get me wrong: fires, floods, meteors, and climate change have had a huge impact. But in the day-to-day world of the genes, the big thing that impacts what works and what doesn't is other species and their genes. If we're actually the super-intelligent species we think we are, we will pay close attention to this fact of evolutionary life when we choose to accept or reject various technologies. Choices based on I, me, now can have devastating effects on the species evolving around us, and that in turn can have devastating effects on us because it undermines our life support systems.

For example, we humans tend to think we're the only species that grows its food purposefully on farms and stores it for future use—we exert huge selective pressures on our crop plants and animals as well as on the species that live in the places where we grow our food as we carry out this biologically ubiquitous activity of food getting[31]. Actually humans weren't the first to farm and we aren't alone in having a huge impact from our actions, but we are alone in being smart enough to realize we have been ignoring the impacts of evolution on our efforts. Other species with less intelligence just live by the rules of nature automatically.

The attine ants were farming for five million years before humans ever started even thinking about it. Above ground, these ants are really destructive—they are leaf cutters and can defoliate a whole area of forest in no time. The foraging ants carry pieces of leaves deep below ground and use them to feed a fungus that their underground farm workers cultivate for food for the whole colony. Some researchers wondered why the ants didn't seem to have any weeds in their "gardens"—molds and mildews and other insects or whatever—none were present. Finally it was discovered that the farming ants that live out their lives deep underground also cultivate a colony of bacteria on their bodies. They use an evolving life form as a pesticide to meet the challenges of culturing a living species for food!

We humans aren't quite so clever—our mainstream technologies use dead, static solutions to evolving problems like pests in our food crops or diseases that affect us. We develop pesticides (mostly poisons made from oil feed stocks) to fight off pests that invade our bodies, homes, or our farms and antibiotics (mostly from the compounds that living, evolving bacteria use to protect themselves from one another and to communicate with one another) to control infections and to increase the growth rate of food animals. Each useful compound we find or develop

---

31 Michael Pollan has done an outstanding job of discussing this last point in his books, "The Botany of Desire" and "The Omnivore's Dilemma."

is mass-produced and widely used. Pesticides and herbicides are spread over vast tracts of farm land and squirted from airplanes flying above the landscape. The glitch in the system is that not all of the individuals of our target (and non-target) neighboring species are affected the same way because DNA mutations of the past have accumulated and new ones keep occurring at a steady rate. Some individuals survive the onslaught of poison. From their perspective, the competition for resources has been removed, leaving the survivors to "go forth, multiply, and subdue" whatever is their source of nourishment.

How does this work? In our fields, our petroleum-based toxins kill susceptible pests, but they also kill the other species of insects and spiders that normally prey upon all life stages of the pests we want to get rid of. Remember the food web? Well, we kill almost all of it along with our target species with this ham-fisted technique.

This means the pests who survive our toxins are living in nirvana—they have an unlimited food supply and have been freed from the strictures of disease and predation because the bugs that used to eat them are dead. Similarly, antibiotic-resistant bacteria have become a scourge in hospitals and long-term care facilities because bacteria have maintained the practices of their ancient ancestors—they are a community of survivors actively sharing innovations that help them to survive and reproduce. For example, resistance to a new antibiotic can spread easily and rapidly to other genera of bacteria that may already have resistance to many other types of antibiotics. Remember bacteria are among the most ancient forms of life on earth and they progressed by sharing innovations and that's exactly what they are still doing. These strains of multiple antibiotic-resistant bacteria kill people because our static, mass-produced solutions are no match for any evolving system. Our unsophisticated toxic solutions to evolving problems create new problems that ricochet back to us as local and global problems. Did you notice I used "problems" three times in the last sentence? That's what happens when the solution isn't intelligently tailored to the problem, that is, we haven't used nature's technique of information-rich methods to control problems. Instead we've used brute force rather than information. Brute force invariably comes back to us as a problem.

A great example of this is house flies and DDT, a pesticide that earned a Nobel Prize for its inventor, Paul Miller, in 1948 because it was so effective in controlling malaria. Given how little we knew at the time, it was reasonable to embrace DDT. It could be safely dusted on people's bodies and caused no sickness. It seemed an almost magical solution to serious, health-threatening, insect-borne diseases, and an almost universal solution to pests that bothered us. It was only later that the second shoe dropped.

I grew up in the country across the road from a barn that housed cows in the winter. This barn was a haven for house flies, which are a true scourge for cows and people but a great source of food for song birds. The neighboring farmer started using DDT to control the irritating flies. At first almost all the flies were killed and we were all relieved except, I'm sure, the birds who had lost an important source of food. Within five years, though, the flies were hardly affected by the DDT. They were back in about the same numbers as they were before the start of the DDT spraying. But although their food supply returned, the birds kept declining, along with the frogs and many other species. It took the work of Rachel Carson and her book *Silent Spring* to get us to understand that DDT persists in the food webs of nature. It concentrates in the fat of each species and is passed along up the food chain. You could tell a long, involved, horrible story here about how this temporary fix created all kinds of problems, but the real message is this. A time, species, global perspective on our technological choices must take into account the setting in which we live—earth's living species are constantly evolving in response to the selective forces acting on them. If we want to control our pests and diseases effectively, then we will model our control mechanisms after those of nature—we must choose living, evolving, dynamic solutions rather than dead, static, short-term sledge hammers.

Non-biological Evolution around Us

In addition to the biological evolution occurring in, on, and around us, the setting in which we live is also permeated with evolutionary change. This evolution, including things like changes in atomic composition of the universe (associated with the big bang), mineral composition of earth's rocks (associated with the action of living organisms), cultural change (originally associated with physical progress in the brain's development), and economic change (associated with cultural change), follows principles similar to the mutation and selection of biological evolution. Here's a look at some background on these innovative ideas that are of central importance as we seek a deep time, species, global perspective.

A McArthur Fellow and professor at the University of Michigan, John Henry Holland, who has an interdisciplinary background in Psychology, Electrical Engineering, and Computer Science, first coined the term "complex adaptive systems" as a set of systems studied in a larger context by the loosely connected interdisciplinary study of complex systems. Complex systems can range from nuclear reactions to cell-cell signaling to international economic systems. In particular, a complex adaptive system (CAS) "is a dynamic network of many agents (which may represent atoms, genes, cells, species, individuals, corporations, nations) acting in parallel, constantly acting and reacting to what the other agents are doing. The control of a CAS tends to be highly dispersed and decentralized. If there is to be any coherent behavior in the system, it has to arise from competition and cooperation among the agents themselves.

The overall behavior of the system is the result of a huge number of decisions (actions) made every moment by many individual agents."[32] Notice this is the mechanism of biological evolution expanded out to include everything from cosmic systems to human systems of all sorts. As a matter of fact, any system made up of elements that can reproduce themselves (either by physical reproduction or by mimicry) and change (either in physical structure or in meaning) will be modified toward an ever increasing share of whatever elements function best in that particular setting. In short, every system of this sort changes in response to the raw material it has available and to the selective pressures put on the situation. Subconsciously, most of us recognize this inevitable process is continuously occurring around us; periodically I've heard media people speak of the evolution of marketing techniques or the evolution of another non-biological system.

Here it is in our faces—a connection that ranges from atoms to our genes to our consciousness to our culture to our cultural outcomes like the economy. All of these systems evolve and all produce emergent outcomes (new features) that result from a multitude of selective decisions between competing

---

32 M. Mitchell Waldrop ,"Complexity: The Emerging Science at the Edge of Order and Chaos" , 1992, Touchstone, Simon and Schuster

and cooperating elements. Each level is influenced from above and below and each level has its own emergent properties that increase in complexity through time by evolution between cooperating and competing elements. We live embedded in an evolving universe and each of our actions is a selective pressure at many different levels of the systems that underpin our lives. See Figure 1.

My use of the word evolution in connection with complex adaptive systems goes to the Latin origin of the word, which means unrolling or change. We commonly use evolution in the sense of constant change as in "the evolution of language" or the "evolution of music" or "the evolution of the universe" or "the evolution of culture". Whether we like it or not, we live in a series of embedded levels of organization all of which are evolving in this sense. That means our choices, our lives, our behaviors are elements in various complex adaptive systems whose direction of change are impacted by us. Heady concept, isn't it. Billions of I, me, now decisions do have an impact far beyond ourselves. Without understanding this, how can we make truly ethical decisions?

To convince you that what I'm saying is true, let me give you a few examples from different levels of organization. Let's start with the bottom three levels of organization in Figure 1. You wouldn't think that the presence of life on earth would have changed how atoms interact with one another to form minerals, but it does. I present this example because it illustrates how each level of organization is connected to, and dependent upon, the levels above and below it. And each level of organization can manifest itself with attributes (emergent properties) that cannot be caused solely by the levels below or above.

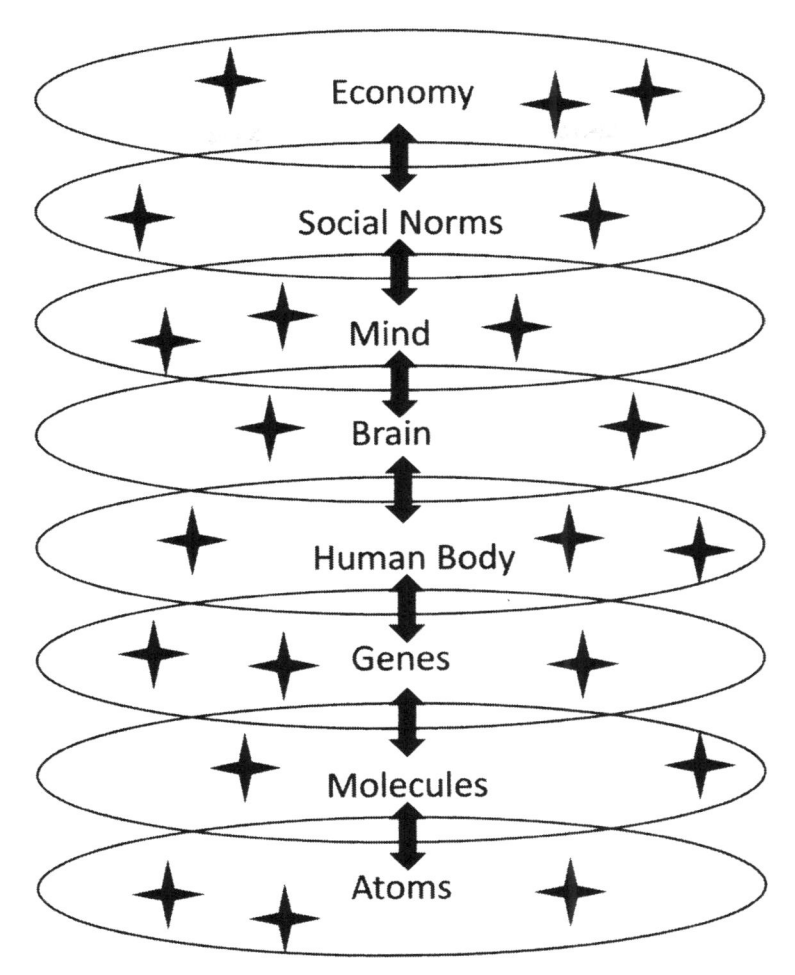

Figure 1. Representation of the levels of organization underpinning humans and their activities. The levels of organization associated with non-human species are left out for clarity. The double-ended arrows indicate interactions between levels and the stylized stars indicate emergent properties unique to each level of organization.

Minerals each have a particular chemical formula as well as a particular crystalline structure. Through time the number of different kinds of minerals has increased dramatically on earth from the 60 found in early meteorites to modern times, when more than 4,300 types have been described. Many of these modern minerals cannot form without the action of living organisms; their structure and beauty are a treat to behold. Here then is a connection between atoms, minerals, and living organisms—each level of organization has its own elements with emergent properties, but the emergent properties are influenced by levels above and below. The whole system has a history of change (evolution) through time toward increased complexity. We can go higher in the levels of organization for more examples.

Our thinking, self-aware, conscious minds are not something readily explained by the biological process of evolution although they clearly rest on a foundation of features like our large brains that evolved though biological evolution. But complete descriptions of the cells of our brains don't explain our ability to be conscious. We can't be conscious without our brain, but our brain's cells aren't enough to explain all the features of our awareness. This is an example of an emergent property; brains can exist without self-awareness but human brains are among those that have this increased feature of complexity. It's our self-awareness and ability to think that puts an ethical responsibility on each of us to reach out for a time, species, global perspective and to learn how to think green.

Speedy Evolution in Culture

We're now at the top two levels of organization shown in Figure 1, culture and the economy. And, yes, each of these things is also a part of the natural world; both are connected to nature through the evolving levels of organization. Culture depends on the interaction of our minds and our minds depend upon the interaction of the neurons in our brains, which of course are a product of biological natural selection. Also, the emergent properties of human culture and economy have a profound impact on all living species as well as on earth's life support systems.

Cultural evolution is possibly the fastest evolution impacting humans today. It is different from biological evolution because the competing and collaborating elements functioning in cultural evolution can spread by learning to genetically unrelated individuals—essentially inheritance of acquired characteristics. Genes can't do that except in the gene-sharing communities of viruses and bacteria—until the advent of genetic engineering, genes in organisms like us could spread only by parent-offspring connections through the generations. This means biological evolution can't change things faster than the reproductive cycle of the organism under selection. Cultural evolution, on the other hand, can result in rapid changes that can turn on a dime in response to the selective pressures occurring at that time and place, the selective pressures essentially being our choices and behavior at any given point in time. This point is the great warning of our current environmental, social, and economic problems, but it is also the greatest hope for the solutions to these problems.

Using GDP to Measure Economy Impacts Economic Evolution

Our large thinking brains and culture naturally give rise to an economy. As far as we know, humans are the only species that use symbols of symbols of symbols to interact with one another. Money is a great example. In our history, the economy must have started by direct exchanges of useful goods between individuals. "I'll give you some of my dried meat in exchange for flint to make knives to cut meat." Then in place of real goods symbols started to be used—rare shells, coins, minerals, precious metals, and things like that were used as symbols of material items—it is so much easier to carry some coins instead of a whole mammoth carcass. Then, we used checks to symbolize money and numbers on paper to symbolize our wealth. Then credit cards, digital exchanges, and now "wealth" flies around the world bouncing off satellites and into computers. Fortunes are made and lost in minutes in the stock exchanges and in speculation on the future value of goods. Symbol upon symbol upon symbol is the hallmark of humans and human culture.

This is a system that evolved to its current state rapidly, much faster than would occur in biological evolution. Our current culture and economy had to have evolved under selective pressures that ultimately come from the people who are a part of the system, but our economy in particular also has emergent properties that seem to run our lives. Because of this, it is worth our time to take a careful look at how people get information about the economy because this information impacts the choices people make. The emergent properties of our economy have a profound impact on our lives in ways ranging from the food we can buy to health care to our ability to make a living. Our economy is of paramount importance to our well being but the ways we measure and manage our economy aren't value-neutral—they have long-lasting effects that can set us on course for serious problems.

A case in point is GDP, which is a measure of a country's output of goods and services. Around the world, we measure the success of countries by the size of their GDP. But this measure is inadequate to measure our well being because it doesn't count the unpaid work of volunteers, child or elderly care givers, or a multitude of other beneficial activities common to human behavior and community well being. GDP does count many things as positives that actually are a measure of decrease in well being: legal fees, medical costs, repairing property after catastrophes like Hurricane Katrina, resource depletion, medical care associated with the health impacts of pollution, car wrecks, and bridge collapses, to name a few. The economists would have us believe that we must all buy into the growth of our GDP as a valid measure of our increased well being. The logic is if there is increased economic output, there are also increased opportunities for jobs to support us and innovations for us to buy. But because the economy is an evolving system, it would be better to measure the health of our economy using a measure that subtracts out things like the money involved in treating cancer, especially if that cancer is caused by pollution put out by something like a regional coal plant.

By giving up the myth that all economic activity is value-neutral, we could more clearly see the impact of the emergent properties our economic policies are causing. Our current system obscures serious problems in a froth of "economic growth" language and sets us up for making countless bad decisions. Right now we are essentially using "gross profits" rather than "net profits" to determine if our economic actions are viable—we all know that isn't any measure of stability for a business because if the expenses are higher than the gross profits, soon there will be a bankruptcy.

Fortunately there's a great alternative to measure economic activity that has already been developed called the Genuine Progress Indicator (GPI[33]). This measure has both pluses and minuses in its calculations. The work of a stay-at-home parent is added to the indicator at the rate it would cost to replace the work and the cost of lost leisure time. Defensive expenditures such as military tanks are subtracted from the indicator. This makes sense because we could make better policy decisions if we were using an accurate measure of how we are doing in terms of human well being.

Right now I'm living in Michigan, a state that has been in a severe recession for about seven years. Some say this is just a product of free trade and the misled American automotive industry. Others say it was the leading edge of the economic downturn that hit the entire U.S. and subsequently the globe in 2008. Regardless, because the Michigan economy is in such bad shape, there are many articles in the newspaper about how U.S. states compare with one another on standard economic measures. I was appalled one day to see that Louisiana's economy was booming according to the reported measures.

In 2005 Hurricane Katrina hit the southern coast of Louisiana as a mega storm pumped up on steroids provided by ocean heat energy caused by global warming. New Orleans was sitting there as a pocket of extreme poverty and extreme wealth, with the poor living in low areas subject to flooding when the badly maintained levees failed. This calamity was "unexpected" even though the disaster was accurately modeled ahead of time and the weaknesses in the system were well known. However, it

---

33 "Redefining Progress: The Nature of Economics". http://www.rprogress.org/sustainability_indicators/genuine_progress_indicator.htm

wasn't considered economically worthwhile to fix the situation so New Orleans, and especially the poor of New Orleans, sat there like sitting ducks in the line of fire. Finally a predictable disaster struck, everyone was shocked, and our social systems that are supposedly ready to respond to disasters broke down. Then there was a nation-wide upsurge of rage over the failures and finally the rebuilding began.

Rebuilding New Orleans and the rest of the Louisiana has brought massive "economic growth" to the state. But what is that economic growth? Is it building something of value that can be used to increase the wealth and well-being of the people paying the bill? No. Instead, that economic growth is actually a defensive expenditure we must pay to clean up the disaster because we don't commonly see metrics that allow us to demand corrective action. What I find most insulting as a taxpayer and as a citizen of the U.S. is that a benchmark like the GPI would have indicated long ago that the misfortunes of New Orleans were a disaster waiting to happen. Investing in the people and the infrastructure there would have been so much cheaper than fixing the wrecked lives and drowned city afterwards.

We must understand and accept that the measures we use to track our economy are moral issues. Each piece of information places a selective pressure on the complex adaptive system that is our economy. It, like all complex adaptive systems, will evolve in response to our choices and behavior but we can't choose in ethical ways when our mainstream metrics are deceptive. With our current metrics, we have essentially counted the financial results of old, infirm people being left to drown in nursing homes in New Orleans as an economic plus—an unbelievably immoral way for us to behave individually and collectively. It is astonishingly unproductive to count the public expense of repairing preventable damage as economic growth—and it is not value neutral to ignore pockets of severe poverty and ignorance in our society. The metrics we use to measure our success have an ethical component because they hide or draw our attention to selective pressures working on the evolution of important aspects of our culture like the economy.

Ethics and Science Work Together

Imagine you are a member of what seems to be the most intelligent species that has ever lived. Your species is intelligent enough to test theories about the origins of the universe that is your home. And your species is clever enough to decode the system that all living organisms use to produce more of their own kind as well as to understand how food and energy flows through living systems. Your species is also adaptable enough to prosper in nearly every spot on its home planet plus carry life support systems into space. And maybe most important, your species is intelligent enough to realize that its home planet is unique in the vast universe. Why on earth would an intelligent species take a sledge hammer to the critical infrastructure that supports its ability to live? The only acceptable ethical behavior is one in which we come of age in our consumption, political, and social choices. The first step to that is a deep recognition of how natural systems function so we can safely take our place us as rightful citizens of this universe.

If we place ourselves firmly as a part of nature and if we recognize clearly how our cosmos and planet are designed, then we will respect that knowledge and start noticing and playing by the rules of this place.

There is nothing in the rules that says we have to live with pests and there's nothing in the rules that says we must die of preventable diseases. There's also nothing in the rules that says I have to allow a mouse to move into my home. There's plenty in the rules that give us clues to how we can most effectively manipulate our environment to keep ourselves and our children healthy, prosperous, and happy. Wouldn't our spiritual lives be fuller if we quit acting and feeling like outsiders in our physical home? Wouldn't the rest of the species on this planet be better off if we quit spreading toxins indiscriminately around the globe? Wouldn't we be better off if we started using the process of evolution to our advantage rather than being continuously blindsided by unintended, but predictable, consequences of our ignorant, static, ham-fisted actions? Wouldn't we all be better off if we recognized our personal power over the evolutionary trends in our culture and economy?

If you still aren't convinced about the ongoing fact of evolution in our daily lives, it makes no difference to its reality. Kusum, a character in F Paul Wilson's Repairman Jack novels, was from the Middle East and deeply religious. Kusum had been setting Rakoshi—vicious, half-mystical, sadistic creatures—loose on the world because of his karma. Repairman Jack was trying to fix the situation. In their final show down, Kusum says to Jack: "Your believing or not believing in karma has no effect on its existence, nor on its consequences to you. Just as a refusal to believe in the ocean would not prevent you from drowning." Old beliefs that humans are above and separate from nature won't protect us from our actions. Evolution impacts all of us regardless of whether we believe it is working or not. In short, dogma, devoutness, and self-righteousness in science or in religion won't keep us from "drowning" in our errors as a species.

Important Concepts of this Chapter

Humans and the species around us share deep similarities because we share a common ancestry from the processes of evolution that have worked on all living things during the history of the earth and continue to work on all living things, including humans, now.

When we pit the great ideas of evolution and creation against one another, we rob ourselves of the synergism that comes when we mobilize our moral/spiritual lives simultaneously with our scientific endeavors. A close collaboration of the two leads us to the time, species, global perspective critical to appropriate ethical choices.

Evolution is a simple process based on selective forces that work specifically in a particular time and place. With the advent of molecular techniques, it's possible to closely track the fingerprint of evolutionary forces on DNA making up the genes and genetic control mechanisms of all species.

Evolution is not a process limited to the past. It is a force going on constantly in humans and around us every day. Good ethical decisions necessitate that we find sophisticated, evolving solutions to our problems rather than static, brute-force methods.

Evolution is not a process limited to the biological world. Any system from a collection of atoms to a group of corporations can evolve if there are competing/cooperative elements that can change and spread their influence in a particular time and place. Human choice is a key selective agent in cultural and economic change.

We cannot make ethical social decisions without accurate economic metrics. Metrics like GDP are deceptive and lead to disasters because they ignore actions that increase well being and count defensive costs as economic growth.

To deal effectively with our evolving home on earth, we must develop evolving solutions to our problems.

* * *

Next I argue that ethical decisions don't entail strictly preservationist constraints. All species modify their environments—the ethical pivot point is confining our modifications to the world of possibilities that don't kill us and the rest of living things on earth. To do so, we must be intelligent and moral enough to choose acceptable behaviors and technologies that fit within nature's time-tested rules. Making this shift from nature's rogues to peaceful citizens of this ancient place will ensure our technologies synergize to provide a healthy future for all life on earth.

Chapter 4

All Species Modify Their Environment

It's spring while I write this. The other day I was out cleaning the algae out of my pond so sunlight could reach the water lilies and make them grow. As I was cleaning, some frogs, crayfish, and other pond residents were watching me carefully. I'm sure a part of their watchfulness was defensive; none wanted to become my lunch. But another part of it felt like their chagrin at my lack of acceptance of how the pond was. When I garden, I always get the feeling that the other species consider me to be the one individual of our community who is never satisfied with how things are. I keep messing up their perfectly comfortable pond. On a broader scale, we humans are masters at modifying the world around us to meet our needs. That's not bad. Just as koalas specialize in eating eucalyptus, that's just our specialty as a species.

Orkin Rules

Many people think that humans inevitably must destroy nearly everything they come in contact with in order to live, but there are good reasons we manipulate our environment to make ourselves more comfortable, as the following story taken from my 2003 Christmas letter demonstrates.

"This has been the year of the rabbits, not the Chinese zodiac kind, the kind that live in your back yard. It's a long, sad story, but please stay with me.

I think it started when I moved into this house 7 ½ years ago. Being the environmentally conscious person, I was running my place with no pesticides or herbicides or chemical fertilizers—just natural fertilizers and pest control except for extreme circumstances. After the rabbits this year, I now have bought stock in DOW Chemical and happily use the most toxic chemicals I can find. You're probably thinking that she's obviously post-menopausal and completely berserk, but let me explain.

It only takes two rabbits or mice to breed thousands in no time. Six years of chemical-free living had created a haven for small mammals on my property so by the summer before last it was nearly impossible to grow a garden without having it chewed off by something. I still blame my mother for part of the problem because she gave me a "Welcome to *our* garden" sign. Anyway, by last summer, I had to put up fencing around anything I wanted to grow and started using AC mouse control to try to keep the field mice from eating the entire garden before it could grow more than 1" high. Now what I've told you already is bad, but the story gets worse. After all, a person can live and even prosper in our culture without a garden. They can't live and prosper with what happened next.

So summer passed and I had formed an uneasy alliance with the odd little community on my property. Then in the fall I found a dead rabbit in the middle of the back yard. Apparently the dog had found it too. To make a long story short, I suspect my rodents and rabbits had caught fleas from somewhere and my dog got infected with them—I'm pretty sure about that because now the dog is so old she doesn't walk out of the yard anymore. Well, the dog didn't just get fleas, she got FLEAS! And I didn't notice (yup, it's been a busy year!).

The last weekend in September, I finally noticed something was wrong and took the dog to the vet, who put some stuff on her neck to make her into a 'flea magnet' and also gave her a pill to kill all the fleas on her. Sounds like a great plan, huh? Well, in short order, things were way out of control. The house got infected and even I had fleas. I would be sitting in my office at work and have fleas jump out of my clothes. It got so bad that I started wearing a cat flea collar around each leg and using dog flea shampoo on my hair. By then I was on a weekly treadmill where I would use every spare minute to vacuum, scrub, wash the dog, wash myself, wash the sheets, and so on but none of it did any good. If anything, the flea population was increasing. I hit a low point one night as I was getting ready for bed.

Because the fleas bit me so much, I had trouble getting enough sleep so I had gone through my nightly ritual of vacuuming the floor and the bed, scrubbing myself really well, then spraying some insect repellent on my legs, running to the open bed, examining my legs for any stray fleas, and getting under the covers. Wow. Felt good to get into bed and to be able to rest. I was lying there and reading a bit before sleeping when I felt a flea in my crotch! I jumped out of bed, ran into the bathroom, found the flea and killed it, and then sprayed myself with insect repellent. Oh man, did it burn! I was jumping around the bathroom and then got to laughing and decided that it was time for the big guns.

The next day I called Orkin and got them to come and kill the little bastards. I also got them to put out poison to kill the rodents and next summer I'm going to shoot any rabbit I see. So now I'm an ardent supporter of the NRA and toxic chemicals—clearly anything is better than trying to live in a flea-infested house."

Oh yeah. I can clearly empathize with people as they've accepted pesticides onto their farms, exterminators into their houses, and lethal products into their homes. I was reading one of Sue Grafton's books the other day, one about private investigator Kindsey Millhone. Kindsey is searching a house to find any potential source of poison that might have been used to murder a victim and comments, "The storage area under the sink turned out to be a rich lode of toxic substances. It was sobering to realize that the average housewife spends her days knee-deep in death."[34]

In our struggle to fight off the incursions of fleas, mice, fungi, bacteria, aphids, mosquitoes, roaches, and the like, we have accepted a whole array of broad-spectrum, deadly substances into our lives. It would be easy to say that humans are a horrible, unnatural species because we use these types of techniques to kill any species that comes within squirting distance. Not only that, humans cut down trees, pave roads, build back hoes to reshape the landscape, shoot rockets and satellites into space, and send submarines to the bottom of the oceans. We use machines and fossil fuels to get work done and to keep ourselves warm in winter and cool in summer. It seems there is no limit to what humans are willing and able to do so there's not a square inch of earth left that we haven't touched with our chemical pollution, extraction of resources, and search for wealth. Those who see and care about the loss of "what-used-to-be" turn on their own species. The rhetoric of many environmentalists casts humans as the enemy, as a destructive parasite that has infested earth and is killing it.

---

34 "I is for Innocent" by Sue Grafton, Henry Holt and Company, 1992.

The simple truth is that all species modify their environment to meet their needs—humans are no exception. Elephants are difficult to keep on nature preserves because they push down the trees to get at the leaves for food. Small species make huge impacts also. The farming leaf-cutter ants (remember them from the last chapter?) defoliate large patches of rainforest in only a few days. Termites live in social colonies in sub-Saharan Africa—there's no vegetation left for large swathes around their huge mounds. Field mice defoliate the area around their homes to make their grass-ball nests. And moles tunnel all over the place in search of their insect food—their tunnels provide homes to gophers and other critters. In the mountains of Colorado, you can see whole meadows formed from the dam-building activities of beavers, which obviously clear cut acres of aspens for materials to build their dams. A whole suite of species moves in after the beavers, including trout, frogs, birds, trees and various other kinds of plants, insects, fungi, and mosses. Bacteria produce and excrete toxins into their environment to keep other bacteria away and many plants produce chemicals that keep other plants from growing too close and competing for water, light, and soil minerals. So if all species are modifying their environments, what's so special and awful about what humans are doing?

The quick answer is "Nothing" if you don't mind accepting the unintended consequences of our actions—things like global warming, depletion of ocean fisheries, extinction of species, hormone-mimicking chemicals, acid rain, peak oil, and premature deaths. The problem is not that we modify our environment with our activities. The problem is the slick of death and destruction we leave behind—we're living in what might be rightly called "The Age of Bad Technology," a set of habits we've adopted unthinkingly from our recent history.

The Origin of American Bad Habits

The "time" part of our time, species, global perspective is instructive here. Our industrialized civilization has developed and come of age in the past 200 years. The first 100 years, the 1800s, were a time of innovation and a time of such abundant natural resources, few people, and pristine life support systems on earth that we thought there were no limits of any kind; Ehrlich's IPAT equation (Chapter 1) would predict the environmental impacts would be small in this setting. At the time, our knowledge of natural systems was rudimentary—it didn't seem to occur to anyone of influence that the massive trees blanketing the U.S. weren't able to re-grow at the pace they were being cut and that these trees were a source of wealth. Plus most didn't want them to re-grow because the ancient primary forests were seen as a barrier to progress, a barrier that had to be cleared in order for the development of civilization. Historically, there was a general attitude that our civilization had to lay claim to the land that was being wasted by Native Americans and the species they used for their way of life.

Then, between 1900 and 1950, we Americans really got rolling in our development. During this span of time, our population increased by 100%, mining increased by 500%, oil production increased by 5,600%, electricity consumption increased by 6,000%, and motorized vehicles increased by 600,000%[35]. In the first half of the 20th century, American civilization used up more non-renewable resources than all

---

35 "Earth and the American Dream", movie, 1992, Bill Couturie

humans had used in all of human history! In addition, during this period our culture developed some paradigms—key deeply held assumptions—which we still hold today.

We Americans have great confidence that technology will solve any problem that arises. Today, in the face of global warming, many are working to develop "clean coal" technologies and thinking about fertilizing the ocean to stimulate the ocean plankton to soak up the excess CO_2 given off from our use of fossil fuels. We also have great confidence in mass production, a technique that works best when things produced are carbon copies of one another. We use monocultures in our food production, build tract houses designed in Texas and for a Texas climate all over the U.S., and drive various brands of cars that all are made on the same basic plan and produced in assembly lines. The FDA has recently approved producing cows through cloning; essentially carbon copies of cows that are raised on carbon copy diets spending their lives in carbon-copy feed lots as though cows can't make enough calves on their own to meet our needs. Tangential to this, we are deeply convinced that economic prosperity can only be based on manufacturing and selling vast quantities of our mass-produced goods to others around the world.

Additionally, and possibly most damaging to our spiritual lives, is the belief that we have a duty to consume because we all know the health of our economy is tied to consumption and our well-being is tied to a strong economy. Consumption, as discussed in Chapter 2, is so important to our culture that stock indexes are reported in all our media. We have developed rituals associated with consumption—Christmas excesses and bridal and baby showers are examples. We are constantly referred to as consumers with lifestyles. We are obligated to buy even if we don't have money. Easy credit is given to all who are willing to accept it, and those who overspend become slaves to making minimum payments on their credit cards. I don't know about you, but I reject the bondage of minimum monthly payments and the corresponding "lifestyle." I am a human, a person with a life. I do not feel obligated to buy things I do not want or need in order to keep the economy strong.

Finally, we are convinced the economy and the environment are forces in opposition to one another—if we do something to benefit the economy it obviously must hurt the environment and vice versa. This is so deeply embedded in our psyches it is difficult for people to see any alternatives. One night I was a guest at the home of some middle-aged people whose father was also present. The father had made a fortune in the automobile industry, supplying parts for cars. The other people present were also deeply embedded in successful lives in our civilization. The conversation came around to what I did for a living, and I explained I taught college-level courses in evolution and environmental ethics. Immediately, the economy versus the environment paradigm reared its ugly and predictable head. It was obvious these people thought all those who cared about the environment were tree huggers who were out to destroy the American economy in order to save some owls or an obscure fish. After giving several examples of businesses that had actually increased their profit by helping the environment, I finally posed a hypothetical situation to the grandfather. "If you had a choice of two things, one that helped the environment and increased your profit margin or one that hurt the environment and increased your profit margin less than the first choice, which would you choose?" He answered that he would have to find a compromise between the two!

Paradigms blind us to new possibilities and bind us to old, wasteful frames. For example, the inventor of the battery-driven watches we all wear today first took his invention to the Swiss watchmakers who were, pardon the pun, the gold standard in watch making at that time. The inventor was sent away—in the minds of the Swiss watchmakers, watches simply weren't made that way. Obviously they were wrong, and it cost them their livelihoods. To move our civilization from childhood into its adulthood, we must reject the limitations of all of these paradigms on our thinking. These paradigms from the childhood of our culture fly in the face of nature's operational rules covered in Chapter 5: waste equals food, nature runs on current solar income and depends on diversity, and nature uses information rather than brute force to solve problems. Because our foundational assumptions have been in opposition to our home, we have created all sorts of problems for ourselves.

By the 1960s, our industrialized activities had created such an ecological disaster that sensitive individuals started to realize our lives were in real jeopardy. Rachel Carson, the environmental pioneer already mentioned in a previous chapter, presented strong scientific evidence in *Silent Spring* that explained how industrial chemicals were getting into the soil and water and were poisoning our neighboring species—worms, frogs, and birds—and were a severe threat to our children and ourselves. Her work led to the first earth day in 1970, when there was an increased consciousness in our culture that we had the power to kill ourselves with our activities and to alter the world. A heady concept, isn't it?

Spreading Success and Spreading Problems

There's no way around it. In the short-term view, our civilization has been amazingly successful. The American free enterprise model along with our toolkit of technological fixes, uniform mass production, obligate consumption, and economic supremacy are currently spreading rapidly around the world, and our power to alter the world has magnified each step of the way. We've been to the moon, to the deepest reaches of the ocean, have satellites orbiting the earth so we can watch our enemies and communicate more freely, and our soft drinks and music reach almost every corner of the globe.

During one of my research trips to Madagascar, I was in a small restaurant in a tiny town with my crew of Malagasy employees. The people in Madagascar speak French and their native language, Malagasy. I don't speak French so I had studied Malagasy for six months before this trip where my purpose was to observe firsthand the world's poster child of unrestrained population growth causing the extinction of one of the most unique and diverse ecologies in the world. I got off the plane after more than 24 hours of travel. The customs agent asked me a question and I completely forgot that I spoke any language at all. The first familiar word I heard was "hello" from a tall, attractive Malagasy man named Rakotoson. He took me to a hotel and I arranged several weeks of tours with him as my guide.

Rakotoson didn't speak much more English than "Hello," so I didn't realize it at the time but when I hired him, I also hired a cook, a pirogue driver, a car and a car driver, a mechanic for the car, and a variety of other local guides and employees. Their normal Malagasy way of dealing with customers is to provide them with nice meals and accommodations while all the employees sleep in and under the car or next to a campfire without bedding. As a middle class American I didn't find this acceptable, so I made it clear that wherever I ate or slept, all of my employees needed to have the same accommodations. So we all ended up in a tiny restaurant that normally served only local people. Water was available on the tables—one used coke bottle with a glass over the top. If you wanted water, you poured some into the glass, drank it, and replaced the glass for the next person.

The native food in Madagascar at all meals is "vary sy loake," a big plate of rice with a little bit of vegetables or meat on the side. In this restaurant, all of my employees were served promptly with their rice and a tiny little fish each. The custom there is to eat the fish completely, bones, eyes, and all. Well, I sat and sat waiting for my food, hungry but a bit wary of having to eat a whole fish with my rice. Eventually they brought me my food, with a fish at least four times the size of anyone else's fish. What a commentary on how our culture of consumption has convinced the rest of the world what we expect, even in the poorest of places!

The size of my fish was embarrassing, but worse yet was the bits and pieces of our culture that had reached this out-of-the-way place. There was a huge Bud Light sign on the wall and music was blaring from a boom box running on batteries—there was no electricity. I don't know the name of the artists, but in English they were singing "Do you want to have sex on the beach?" I'm not sure the Malagasy people would have been willing to listen to that music if they had understood the words. The spread of American culture and the rapid economic development of formerly poor countries are also spreading the American way of life and our I, me, now worldview.

Remember the arithmetic of Ehrlich and Brower?

Impact = Population × Affluence × Technology.
The world we have lived in up to now has supported a few wealthy people living the American lifestyle, which is resource-intensive and based on a series of environmentally damaging technologies. Wealthy people have created problems like climate change, loss of biodiversity in the far reaches of the globe, and other global environmental problems that promise to create global catastrophe if we don't change our ways quickly. Now we are faced with millions (billions) of other people, formerly poor and living lives using few resources and benign technologies like subsistence farming, striving toward an American way of life. So the Ehrlich and Brower equation is on steroids as I write this. More and more people are becoming "Americans" with American consumption patterns. But are we spreading the right things?

Design Goals and Unintended Consequences

As we saw in Chapter 3, cultures evolve just like bacteria and elephants and all the rest of living species. Remember cultural variations subjected to positive selection (chosen by more people) become more prevalent and those selected against become less prevalent. Because we are so wealthy and successful *in this time and place*, right now there is rapid positive selection for our culture around the entire earth. Because there are such obvious cultural, economic, and environmental problems associated with America's apparent success, it is worth our time to look carefully at the unintended consequences our culture has produced. A good way to understand the magnitude of unintended consequences and how they come about comes from the work of William McDonough, an architect from the University of Virginia and a businessman. An obvious question would be how an architect would have anything of value to say about something as weighty as the global impact of our culture.

But think about it. Architects are in the business of design, and as McDonough puts it, "Design is the first signal of human intention." So to help his students understand the problems with our current version of an industrial culture, McDonough gives them an assignment to retroactively design the modern industrial world that underpins the American Dream. McDonough's assignment goes like this:

> "I'd be asking you to design a system that puts billions of pounds of highly toxic material into your soil, air, and water. It measures prosperity by how much natural capital you can dig up, bury, burn, or otherwise destroy. It measures progress by your number of smokestacks and, if you're especially proud, you put your names on them. It measures productivity by how few people are working. It requires thousands of complex regulations to keep you from killing each other too quickly and destroys biological and cultural diversity at every turn and, while you're at it, do a few things that are so highly toxic, like nuclear isotopes, that they require thousands of generations...[of]...constant vigilance while living in terror. The operative design principle is if brute force won't work, use more of it." [36]

With this perspective on the results of our industrial culture, we must ask ourselves what we have been accomplishing.

---

36 "The Next Industrial Revolution: William McDonough, Michael Braungart and the Birth of the Sustainable Economy," movie, 2001, Earthome productions.

Certainly not what billions of individual I, me, now decisions intend. We've been making choices that improve our quality of life by promoting the health of our economy in ways measured by metrics like the GNP. If you look at our success from McDonough's perspective, we haven't been very successful because our unintended consequences are now threatening earth's life support systems that underpin our very existence.

If we put the ideas together that we've been talking about so far in this book, then it is clear we must quickly develop a new set of design goals for our activities. These new design goals must free Americans from obligate consumption while improving the quality of our lives. This means quality products must be available on the market that are produced without spewing pollutants and without destroying non-renewable natural resources and they must meet real needs of people, not just the profit needs of corporations. Our new design rules must function to strengthen and improve earth's life support systems for humans as well as other non-human species. Without that, we have set ourselves on a course toward deprivation, disease, and economic collapse. We can modify our environment to meet our needs, but we must think of ourselves as a part of nature and subject to its rules. A key part of that is to recognize the evolving nature of all of the levels of organization around us from atoms to ecosystems and economy. We must be vigilant of selective pressures if we are to succeed in accomplishing our past success with new design goals that utilize a time, species, global perspective. Some of this work is already started.

McDonough is working very profitably toward a new industrial revolution, one that sees itself inside of nature and follows the principles apparent in ecosystems that have stood the test of time (more on this in chapter 11). Another way of saying this is if we do things in an intelligent manner, we can have our cake and eat it too. We can manufacture products, produce our food, and keep our homes comfortable (and flea-free!) without killing nature's life support systems or ourselves or decimating other cultures. But it's going to take more brains than brawn to make the transition. We must recognize our actions are occurring in the context of evolution, ours and of the species around us, and that our actions must fit within nature's operational rules covered in the next chapter.

Fortunately there are many people in commerce with a time, species, global perspective who see this clearly. We'll cover this is more detail in Chapter 11, but for now it's sufficient to say that there is a whole sustainable business movement in the US and globally where corporate leaders are taking responsibility for the social and environmental impacts of their activities as well as the economic impact of their corporate activities. These pioneers are recognizing they must value human capital along with natural capital as they go about making non-toxic products in ways that meet the real needs of people. Once again we can celebrate the creativity and intelligence of our fellow humans!

Clinging to Our Old Ways

What happens if we cling to the solutions and paradigms of the old industrial revolution? The messages around us are clear. Our civilization won't persist for the long term because we will have run out of resources, overshot the carrying capacity of the planet within the framework of our material needs, and spewed out so much waste that the stability of earth's climate and the ability of species, including humans, to persist is compromised. This won't be the first time this has happened to human civilizations[37] and it won't be the first time the activities of a species has led to widespread extinction. In geologic time, most of the major extinctions have been caused by meteors hitting the earth or by climate change driven by new connections between oceans or by mountain uplift. The extinction we are causing and the catastrophe our civilization is facing is due to our own activities that trample the principles of nature. I know of only one other instance in the geologic record where living species caused the mass extinction of the majority of life on earth. This happened early in earth's history.

As you recall from Chapter 1, the earth-moon system originated about 4.6 billion years ago. By about 3.6 billion years ago, early bacteria had evolved and were prospering in an atmosphere that lacked oxygen. In fact, oxygen was poisonous to most of these early life forms. We know they existed and what they were like because some still exist today, living on the fringes of a world that has been poisoned for them by the presence of oxygen in the air. Then, as now, life is based on continual change and there's always something to spoil the party.

In this case, it was the development of a new group of organisms that could make use of the CO_2 given off by the old bacteria—in other words, the development of photosynthesis that uses the CO_2 waste as food to make high energy organic molecules with energy from the sun. The waste from this process of photosynthesis is the life blood of our lives, oxygen in the atmosphere. It is possible to track the increase of oxygen in the atmosphere because iron rusts in the presence of oxygen—as oxygen increased in the atmosphere, the minerals

---

[37] Jared Diamond, "Collapse: How Societies Choose to Fail or Succeed", 2005, Viking

in rocks containing iron that were formed at various geologic ages become more and more oxidized (rusted). Here then, was a global catastrophe for the old bacteria, but a whole new world of opportunities for new life forms. Our ancient ancestors were the beneficiaries of this changed atmosphere. Without abundant oxygen, our highly complex and energy-intensive bodies couldn't exist.

If there is a message here, it's this. Just like the ancient early photosynthesizers, humans are not capable of wiping life off of the face of the earth. Even a nuclear war wouldn't accomplish that. We are capable of changing this place though so it isn't livable for ourselves or for many of the species that share the earth with us. But as Dr. Malcom, the mathematician in the movie "Jurassic Park" says when they discover the revived all-female dinosaurs on their remote island are reproducing, "Life will find a way through." This is the power of the evolutionary system in place on the earth and in the universe.

Ethical behavior now, with our depth of scientific knowledge, demands that we grow quickly out of the paradigms of our civilization's childhood and adopt a new system of assumptions. We must learn from the lessons written in the geologic record, in the systems around us, and in the history of civilizations on earth. Our actions must be guided by the principles of nature so our thousands of individual "good" decisions will add up to a prosperous future rather than a massive disaster. There's no reason for disaster to occur if we are intelligent enough to learn from the time-tested rules of nature. Yes, our activities will modify our environment. But as long as we leave a slick of health and abundance behind us, the rest of nature will evolve along with us and synergize with us just as occurs in any mature ecosystem.

Important Concepts of this Chapter

All species modify their environment. Humans are no exception.

The damaging technologies we are currently using are a product of our ignorance as our civilization was developing. As our population has grown, the paradigms we've developed are no longer functional. Technology can't solve every problem, the purpose of our lives is not consumption, and it is archaic thinking that pits the environment against the economy.

We are exporting the damaging practices of our civilization to every corner of the globe whether we realize it or not.

To be ethical, we must be aware of our design goals in our activities and we must abide by the time-tested principles that govern nature: waste equals food, run off of current solar income, depend on diversity, and use information rather than brute force to solve problems.

This isn't the first time a species has caused a catastrophic change in the course of life on earth but this is the first time the species involved is intelligent and can see the writing on the wall that calls for particular changes that will avert the coming disaster.

It is our ethical responsibility to pay attention to the warning signs.

* * *

In order to make the maximum use out of the resources present on this ancient earth and in order to increase the abundance of resources we need to make ourselves comfortable and healthy, it's vital that we recognize the unwritten rules that have given rise to the abundance present on earth. The abundance of which I speak includes the stored resources that enrich this planet, the currently underutilized resources that could improve our quality of life, the 1.8 million species with whom we share this planet, and the network of interrelationships that allow all species (including humans) to make a living. We can best do this by becoming students of what works for the rest of living organisms.

Chapter 5

Nature's Operational Rules

If we want to live a life outside of the onion and proudly retake our place as a part of nature, there are a lot of things we have to know. It isn't just as simple as going green because the marketers know Americans are becoming increasingly concerned about the environment so the new handles used to capture our attention frequently entail some kind of green wash over the same old ineffective and destructive products. To actually choose our basic necessities using a time, species, global perspective requires that we take a careful look at nature.

The first step in understanding what is really green is to realize that we and all other living species on earth rely on earth to underpin everything we need. To live well, each person must have adequate, clean, pollution-free food, water, air, clothing, and shelter. They must also be able to live in a mentally healthy community where their needs for education, health care, socializing, privacy, mobility, esthetics, employment and spirituality can be met along with the physical needs listed earlier. Put another way, we all rely on earth's life support systems for our food, water, air, weather, social order, and economy. This is non-negotiable. Humans have never invented anything without using the materials and resources available in the visible matter of this universe. And all of what we've "invented" has come by using visible matter in the form of the living and non-living resources present on this planet. It's not easy to become aware of and to gain respect for this profound dependence we have.

For years I've been a part of a Science-Religion study group. We've read and dissected nearly thirty books dealing with the interface between science and religion. One that struck me most forcefully was by B. Alan Wallace, *Choosing Reality: A Buddhist View of Physics and the Mind*[38]. He made the important point that the accuracy of what we can know scientifically is limited by the sensitivity of the instruments we use to measure and collect our data. Most important among those instruments is our minds because if we are unaware of subtle things, we have missed seeing important parts of the data. In this book, he has the reader sit and concentrate on breathing in and out. I did what Wallace suggested and almost fell off my chair because each time I breathed in, the tissue around my nose was momentarily colder and each time I breathed out, it was momentarily warmer. I had been breathing my entire life and had never noticed. Similarly, a time, species, global perspective cannot be gained unless we notice and value our detailed dependence on earth.

One other story is important here to help us keep the power of our intelligence and science in perspective. Doug Kindschi, a mathematician, philosopher, academic dean, former boss of mine, Presbyterian, and the leader of the Science-Religion study group told the following story a couple of times over the years. At a futuristic time, the scientists had finally gotten to the point that they understood all the workings of life. They could build cells from scratch and get them to develop into whole organisms. One scientist went to God and said humans had finally figured

---

38 Wallace, B. Alan. 1996. "Choosing Reality: A Buddhist View of Physics and the Mind". Snow Lion Publications, Ithaca, new York.

out how to make life. The scientist said, "First you start with a handful of dirt..." God broke in and said "Uh, uh. You have to go get your own dirt."

As we look for the real meaning of green in our quest for a better quality of living as Americans, the first thing we must learn to do is to look to the world around us for the basic principles that are built into this place. That's like the dirt in the story above. As we've already discussed earlier in this section, this place is ancient and it is characterized by ever increasing complexity, innovation, pragmatism, and constant change. Underlying this universe's ability to continuously develop is a set of principles. But we must sensitize ourselves to these principles and keep them in mind if we are to be able to correctly interpret and respond ethically to the choices we face as the most privileged people on earth. Just as I became aware of the warm and cold air at my nostrils, we must become aware of earth's rules for living that are written large and small all around us. And we must be always respectful of the limits of our beliefs and data as well as the value of the "dirt", the bounty we have inherited by virtue of this place.

Nature's Principles

It's difficult to find a spokesperson for nature but from my 34 years of postdoctoral scholarship, I've never found a better one than Paul Hawken. In those years I've known countless biologists, geologists, chemists, and other professionals from across the spectrum of academic disciplines and commercial specialties. To me it seems like each academic discipline and each profession is like one person touching a different part of an elephant in great detail but even though many people are studying their piece in great detail almost no one has more than a glimpse of the whole elephant. Paul Hawken is different.

Paul Hawken is a man who has built his life around developing a deep understanding of many facets of our environment, our lives and the social order in which we live. As an environmentalist, business owner, and journalist, he has dedicated his life to changing the relationship between business and the environment and between human and living systems with the purpose of helping to create a more just and sustainable world. Hawken is widely published in papers, articles, op-eds, and books. My favorite Paul Hawken book is *The Ecology of Commerce: A Declaration of Sustainability*, a book I know has changed whole industries in the region where I live. Hawken's work is used in prestigious schools of business around the US and is foundational to a new philosophy of sustainable business practices that are gaining strength rapidly in our culture. To the normal American, this means we can cue in on his ideas to help us discern green-wash from green as we work to rewrite our relationship to the earth.

According to Paul Hawken, three principles govern nature[39]: waste equals food; nature runs off of current solar income; and nature depends on diversity. We'll start with those three principles but because of my 35+ years of broad-based scholarship, I must add one other principle: nature/earth uses information rather than brute force to solve problems. Let's look at each of these principles in enough detail to make sense of them as they relate to us.

Waste Equals Food

First, "waste equals food"—a deceptively simple idea with profound implications—is an idea worthy of the ancient, highly

---

[39] Paul Hawken, "The Ecology of Commerce: A Declaration of Sustainability," 1993, HarperBusiness

evolved organization on earth. Another way of stating this principle is that physical material, both living and non-living, is remade, recycled, and reutilized over and over again in this place. This is true from the subatomic level through the level of stars and beyond. It includes the substance of our bodies as well as water, rocks, trees, mice, and atoms. Waste equals food is foundational to the cycles writ small and large around us.

In biological systems, lots of things are used in huge amounts and many things are "wasted" by each particular species but the waste becomes the food of a plethora of other species and the foundation of whole new communities. For example, dead leaves feed worms but they also feed bacteria and other soil microorganisms that have intricate relationships with one another and with the larger world of the worms. Similar things could be said about the trees, plants, and animals around us or about the community of microorganisms living in the ocean. Ecologists call these relationships a food web if many species are involved and eat similar things or a food chain if only a few species are involved.

Waste equals food also involves relationships between the non-living and living realms. The whole earth depends on biogeochemical cycles to turn the waste of one system into the food of another system. Carbon, phosphorus, nitrogen, and water cycle between the living and non-living components of earth; the waste of one process (e.g. humans breathing) feeds the composition of the atmosphere as well as the chemistry of the oceans and cycles back to the bodies of living organisms. On Earth, non-living things also recycle among themselves. For example, there's a cycle in rocks; rock types are transformed into one another through the processes of erosion, sedimentation, heat and pressure, and melting and cooling and, as they change, they affect the biogeochemical cycles just mentioned as well as the earth's crust (mountain uplift, ocean formation, continental drift, and the topography of our country).

It's as though the theme of this place we live in is waste nothing permanently, use everything in the short or long term, and stick to materials that can be recycled in the context of the current organization that is in place. This is an expansive statement, but let me put it into a familiar context.

In biological terms, all living organisms gain their energy for living by chemically manipulating a confined universe of naturally occurring molecules. Had the frozen accidents of our early history occurred differently, we would have a different set of "naturally occurring molecules" commonly used by biological organisms if they existed. But things are as they are and we live in this universe with the form of life that is here. Strange molecules like trans fats, made by chemists in food labs for particular commercial purposes related to the taste and texture of mass-produced food, are not naturally part of the universe of molecules utilized by living organisms on earth. Therefore our bodies, products of the dominant cycling systems on earth, cannot correctly metabolize foreign molecules like trans fats. This is probably why our bodies tend to pack molecules like trans fats into the linings of our arteries because the shape (chemistry) of the molecule is such that it won't "fit" into the chemistry of our bodies.

There are many things we have been doing in our recent history that don't "fit" into the chemistry of the cycles of earth. Plastics are a prime example. There is no organism on earth that can use plastic as its food source—plastic goes straight through a digestive tract unscathed except for mechanical damage so it short circuits the waste equals food cycles at all levels of earth's organization that involve living organisms. If we are to become truly green, then we must replace materials like plastic and trans fats in our technology with materials that fit within earth's universe of commonly utilized molecules. Because of health considerations, all molecules that become part of our food should only come from the universe of biologically-derived molecules. Manufacturing of consumer goods may be a different story in some cases.

If we must utilize chemist-produced compounds for manufacturing, they should be isolated within a human-generated cycle of recovery and reuse. This idea isn't original with me. It comes from the design-goal thinking of William McDonough and what he calls C2C (Cradle-to-Cradle) thinking in the manufacturing enterprise. Materials used would either be part of a biological recycling chain or of a technological recycling chain. Of course, it won't be easy to become this sophisticated in our activities but unless we adopt goals in line with this waste equals food foundational principle of earth we will continue to evolve our technology into paths that damage ourselves and the stability of earth's life support systems.

This takes us to the ideas of wasting nothing permanently and making sure waste can be utilized in the short or long term to increase the health, wealth, and well being of this place. This is a key place where the model our civilization has developed differs from the waste as food natural order of this place. The waste of the natural order is non-toxic so it can stay in place and eventually some species will utilize the waste as food. Obviously we must keep toxins out of our biological waste stream. We cannot allow toxins like mercury, lead, and arsenic to contaminate our biological waste. And we cannot confine a mixture of biological and technological waste to EPA-approved landfills that are designed to put our waste into a state of suspended animation until failure of the landfill occurs and some later generation must clean up another toxic site in their environment.

Instead our waste should feed either an industrial food chain or it should be composted as quickly as possible to enrich the fertility of the earth. We'll cover more of this thinking in Chapter 11. If we adopt McDonough's C2C thinking, we will be wealthier and healthier. To accomplish this shift, our individual and collective mind set must change from discarding everything we don't want as a mixture of resources we call trash to noticing the resources present in our waste and insisting they be used in a cyclical manner to replenish the supply of our necessities. On a personal level, this sounds like we must become compulsive about our trash but that really isn't the case. Instead we must become extremely selective about what we buy. If we do, the piece of any cycle that occurs within our bodies, homes, and families will take care of itself. I'll discuss this more in Chapter 10.

Nature Runs Off Current Solar Income

The second principle is that "nature runs off of current solar income." In this principle, Paul Hawken is talking about the fact that energy obeys different laws than the atoms and molecules we talked about above. Thus, earth's biotic (living) and abiotic (non-living) systems have adapted to those laws by developing mechanisms that efficiently transfer energy, store energy, and utilize exact amounts of energy to function. To understand energy, we must take a brief look at the laws of thermodynamics.

Energy doesn't cycle. It is present in a limited and stable amount in the universe and it can transfer directly between systems as heat without changing form. Energy also can change form and be transferred in the process. For example, the potential energy stored in the cells of your body is transferred into kinetic energy as you go about your daily activities. Because not every movement is purposeful or productive, some of the energy is lost to entropy (randomness) and because we must have energy to maintain our bodies, some of the energy is lost as heat to our surroundings. I don't know about you, but some days I seem to be short of potential energy and long on entropy and heat. Between the two, it keeps me short of my potential!

Puns aside, the non-cyclical nature of energy and the sun as earth's ultimate source of it is the crux of Hawken's second principle. We can see the non-cyclical nature of energy in biological communities and in our own food supply. Biological communities run on solar energy captured by plants that absorb energy from the sun and make chemical energy, which is stored through the food-producing process called photosynthesis. Photosynthesis makes use of a complex network of light capturing pigments and the precise cellular structure in plants to utilize the sun's energy, carbon dioxide, and water in the leaves plants to produce high-energy organic molecules that provide energy for all the rest of living organisms. Plants, including those from tiny to huge, are the only organisms that can do this so they are called the producers of the community. Every other organism in the community lives by eating and metabolizing the chemical energy stored by plant photosynthesis. Each time the energy enters the body of another organism, only about 10% of it is available as food. For example, rabbits eat plants but only about 10% of the energy eaten by rabbits is available to coyotes when they eat a rabbit. And when the coyote is "eaten" by the decomposers like fungi, earthworms, and bacteria, only about 10% of the coyote's energy is available to the decomposers.

Similarly, when people eat plants more energy (food per acre) is available to them than when people eat beef or chicken or milk or eggs because of the systematic loss of energy to entropy, work, and heat. This loss of energy as it is transformed repeatedly between forms of energy is inevitable because this is how our home in this universe works.

Most Americans are not aware of this hard fact because our ancestors, clever people that they were, discovered that they could "make" energy by burning fuels of various sorts. So now instead of using current solar income our American culture has adapted at this unique time in history to run on cheap oil, which is the fossilized product of photosynthesis that occurred somewhere around a half a billion years ago in the Paleozoic Era. For perspective on this number, a half a billion seconds ago, it was 1991. On average Americans use about three gallons of oil per person every day[40] of this fossilized solar energy. The glitch is that as we release this ancient stored energy, we are also releasing the CO_2 sequestered in the high-energy fossilized plant matter. At the same time we are releasing this extra CO_2 into our modern biogeochemical carbon cycle, we are also decreasing nature's ability to use this "waste as food" because we're cutting forests and blanketing our habitat with turf grass that absorbs much less CO_2 in its photosynthesis than does producer-rich dense forests. This means our modern carbon biogeochemical cycle is overloaded with CO_2 in the atmosphere, a feature that has been correlated with higher global temperatures in the an-

---

40 Tim Appenzeller, "The end of cheap oil," National Geographic, June 2004.

cient history of the earth because a CO_2-rich atmosphere keeps more of the sun's energy inside of the outer atmosphere.

A key global system impacted by the total heat energy present inside of the atmosphere is earth's climate and the weather events that in the long term accumulate to produce climate. Heat that transfers from the equator to the North and South Poles, the movement of the polar jet stream, the temperature of the ocean surface and the ocean depths, and many other factors determine evaporation, wind speed, storms, and precipitation. Climate is a massive, chaotic system and the more energy in it, the more pronounced weather events become. Therefore, the canary in the coal mine for global warming is not the temperature of any particular place on the earth but rather the frequency of record-breaking weather events.

I consider record-breaking weather events to be the best indicator of the stability of the climate system because of results from my studies of ancient climate in south-central Colorado. In the National Geographic-sponsored research I did with the collaboration of noted researchers from around the country the main climate signal that occurred as the climate changed from warm and mild to the rigors of the ice ages was a period of record wet, record dry, record cold, record warm alternating with one another in various combinations. The impact on the plants and animals living in that place was profound. Many went extinct there and many new species moved in from elsewhere. Because we have so many people now and because we are so dependent on our social order, this type of climatic disorder would have a profound impact on our ability to raise food, an ample supply of which is the foundation for a stable social order. This is a primary example of why it is critical that we adapt our society to live on a strict energy budget that is as close to living on current solar income as possible. For the portion that isn't possible, we must adopt the waste as food principle to ensure no excess CO_2 is released into the atmosphere.

Living the way nature does doesn't mean we have to live huddled under a pile of blankets in our cold houses during the winter or sweat our way through heat waves in the summer. Other species have successfully designed (under the strict "market forces" of natural selection) homes that have a constant internal temperature without using any fossil fuel. In particular, termites living in Sub-Saharan Africa build their mounds with an intricate system of ventilation so their home maintains comfortable temperature and moisture levels even in the face of ground surface conditions that would kill the termites. If they can do it, we can too. And if we started using this kind of technology, think how much money we would each save per month on our "energy bill". I like the idea of sending energy bills extinct.

I hope you see the important point of this section. If we were to strive to live on current solar income we would be developing all sorts of new homes, modes of transportation, efficiency mechanisms, and energy-collection devices. We would learn to meet our necessities and luxuries without damaging earth's evolved biogeochemical cycles nor its energy equilibrium. In return, we would have a cleaner environment, more comfort, better health, and more freedom from albatrosses like energy bills.

Nature Depends on Diversity

The third principle of nature is that "nature depends on diversity" rather than monocultures, clones, and mass production. Remember from Chapter 4 how we Americans prospered during our industrialization by adopting mass production of goods that were produced on assembly lines as carbon copies of one another? But that isn't how nature prospers in the long term because the prosperity of nature is driven by the unforgiving market forces of natural selection acting on mutations (innovations) at each local place. Anything that doesn't work *best* in *that time and place* is replaced by what does. Therefore, each local place has its own set of best solutions that fit with the historical, physical, biological, and climatological conditions of that local place.

As we discussed in Chapter 3, evolution functions in and around us every day all day impacting everything from the disease organisms that infect us to the economy that impacts our ability to make a living. Remember the process of evolution is a local phenomenon—mutation occurs in a local individual and the power of selection operates at a local level so what is best in one spot isn't the same as what is most adaptive in another spot. Another way of saying this is that if there were no mutation and if everything were uniform, there would be no evolution, no progress toward complexity or toward stability of interrelated systems. This observation is true at a variety of different levels.

Think about human cultures through time and around the globe. China (but Mongolia is different than Shanghai), Madagascar with its 18 tribes, Kenya, Costa Rica, the Neanderthals, the Incas, the Inuit, the Greeks, the Romans, the Polynesians who spread into the Pacific islands, the southern belles of South Carolina, the surfers of California—all cultures with unique aspects tied to their history, climate, nature around them, and foods available. Think also about broad areas of the globe—desert creatures are vastly different from rainforest species and swamp critters bear little resemblance to mountain species. Think about the people in your own family—closely related to one another but each is distinct. Think also about the species around you in your everyday life—each is distinctly adapted to its own niche in the greater scheme of things.

The work of Professor John Marzluff from the University of Wisconsin is instructive here. Professor Marzluff was wondering if humans could tell individual crows apart. Crows are an intriguing research subject because they are one of the most intelligent species that commonly lives around us. The research protocol called for his graduate students to band crow chicks while they were still in the nest so they would carry a unique marker in addition to their facial shape, bill size, and other cues that might be used to tell the crows apart. Apparently the crow parents were really upset by the intrusion in their nests. After the banding, any time the students who did the work walked out of a building on campus, they were followed by raucous calls from every crow around. Other students weren't bothered by the crows. Obviously the crows recognized individual humans.

But how did they do it? Sight? Smell? Body shape? Intuition? To find out, Professor Marzluff had his students repeat the banding procedure while wearing a Frankenstein mask. After that, the crows would harass any student wearing the mask but they didn't react to a mask of Dick Cheney. Finally, Professor Marzluff had the students try the mask upside down. No reaction from the crows until one crow turned its head sideways, looked hard, and then started his raucous call. The other crows cocked their heads in a similar way and soon started warning everyone this nasty creature was present who messed with their nests.

The flip of the coin, humans telling crows apart is a different story. The crux of the story is that we can't. This research was reported on NPR in July 2009. To test this NPR posted a crow lineup on their website. They first showed a picture of an individual crow and then allowed people to pick the individual crow out of a set of six pictures of crows. People couldn't do any better than random guesses. We simply can't tell crows apart by their faces. Together these two observations form Professor Marzluff's "Crow Paradox"; Crows can tell us apart but we can't tell them apart. Why? Professor Marzluff hypothesizes it's because different humans make a difference in crow lives—some people feed them and other people shoot them—whereas individual crows make no difference in our lives. Natural selection has worked on crows to endow them with the ability to tell people apart but we haven't been subject to this kind of selection. This kind of difference, while entertaining, is also a strong reminder to us that variety within species, between species, within climate zones, between climate zones, within each place on earth, and between each place on earth is an outcome of the history of the evolutionary forces that are working and have worked in the place in the long term.

On earth, diversity is the name of the game. Monocultures, clones, and mass production are antithetical to this principle of nature. Think about our current paradigm of mass production. We have used this model in all aspects of our lives from food to clothing to our homes to the education of our children to the recently crippled global economy. We now have industrialized fishing (Chapter 6), industrialized agriculture (Chapter 7), industrialized housing (aka subdivisions), industrialized education (single-aged student groupings), industrialized economy (Chapter 2), and industrialized social order (groupthink propagated by the media). We have been staving off the rigors of natural selection but the pressure has been steadily increasing on us. Higher and higher percentages of students require special education, global food shortages and widespread food poisonings have occurred in the recent past, and many of us bemoan the direction our society is heading. We all bemoan the impact of our recent economic crisis. This mass-produced, monoculture, industrialized mind set is not working well for us because it leads to systems that are uniform and therefore inherently unstable.

For example, if you're a species co-evolving with humans (Chapter 3), then all you have to do is to hit a mutation that works in the monoculture system. In short order, you can blanket the earth along with the human-designed monoculture. Imagine the repercussions of a pesticide-resistant mold that could thrive on our modern genetically engineered cloned corn that is grown in countries around the world. Imagine you are a plant humans find useful or beautiful. Suddenly, you can blanket the earth with your seeds and bring your pests and diseases with you. The same can be said of social systems.

As mentioned previously, humans in each part of the globe have evolved unique cultures that are adapted to the local conditions in each particular place. This diversity is being replaced by English-speaking, free market economies, American music and video, as well as the ubiquitous influence of transnational corporations as they seek resources (including human labor) and markets for their products. This monocrop of American culture is spreading like wildfire around the planet replacing ecologically viable cultures that have endured the test of time.

Please don't misunderstand me. Change is inevitable on this planet and change always entails the death of the old and the spread of innovation. It's also clear that some cultural practices beg to be sent extinct—female circumcision in northern Africa, dowry killings in India, Jihad in the Middle East, within family murder in New Guinea, and so on—but the loss of diversity in human cultural practices around the world decreases the local adaptation of humans to their particular environment and also impoverishes our species from the standpoint of innovations that may improve our future.

Human-designed social systems co-evolve with other social systems. As we blanket the world with standard ocean shipping practices, things like modern piracy predictably evolve. As we spread US agricultural practices to remote corners of the world, family farms are lost to agricultural aggregates that industrialize food production and narrow the opportunities for people to make a living as well as narrow the varieties of organisms grown for food. Traditionally, the combination of food preparation methods and the foods eaten in each local place was adapted to what could be grown or harvested in that place and prepared in ways that ensured the best health available in that time and place. For example, the ales of old England provided key nutrients to a balanced diet in the Middle Ages. Each loss of local diversity—biological or social—reduces the resilience of the web of relationships present because mass production simplifies every system it touches. Mass production techniques also destroy the complex evolved relationship between the people and their place. The more simple any system, the less stable it is.

As we saw above with biological systems, clones are an evolutionary opportunity for pest species because just one innovation is sufficient to spread the pest and its colleagues everywhere the victim species is the same. Without genetic diversity the victim species is a sitting duck. Without genetic diversity, any small change in the climate or soil is beyond the genetic potential of the clone. Social systems like culture and culture's offshoots like economy are in the same boat.

A global economy based on uniformity is vulnerable to any small perturbation that occurs. Overpriced housing in the US becomes sufficient to send small countries in Europe bankrupt. Like dominos, the elements of the global "food web" in economy topple just like what occurs in biological systems when diversity is lacking.

There's no two ways about it—in this world, lack of diversity is inherently unstable so if we want to reduce our risk we must learn to appreciate and cultivate diversity. Biological, cultural, and economic diversity are the basic resources for innovation in our world because diversity is the raw material sorted by selection to provide adaptations that fit to local conditions in each place. Evolved solutions that fit each local place are resilient to catastrophes that occur and reduce risk in the long term.

Information-rich Solutions

Surely I wouldn't place myself among the intellectual giants of the world like Paul Hawken but as I've studied and read and learned through my long career, I see a key piece missing from Hawken's three principles. That may be because Hawken was targeting mainly business people in his presentations I have read but be that as it may, I must add one more principle to Hawken's list. Nature uses information-rich solutions—it is infinitely subtle and sophisticated especially when compared to human-designed systems.

Think for a minute about human reproduction. Promiscuity aside, most people purposefully choose a mate with whom they want to produce and raise children. Each of us is attracted to some people but not to others even though they may be perfectly nice people. What subtle signals are involved? In an evolutionary sense, the signals are at work to help us choose a person whose genes complement our own for reproduction. Research in this area suggests one factor in our choice is a mate whose immune system differs most widely from our own, a feature that would help our offspring resist disease more effectively and thereby improve their survival. Because people don't come with packaging labels, we must be endowed with subtle talents that are subconscious yet effective for mate selection. This skill we have must depend on sophisticated systems within our bodies that express themselves as emotional attraction to particular people and no attraction to others.

Information-rich solutions don't stop with attracting males and females to one another. Once egg and sperm are united, all of the genetic information needed to produce an entire new human is present in just the tiny fertilized egg. In a healthy uterine environment, this program unfolds step by ancient step to generate something as wonderful and precious as a newborn member of the family. The baby is unique in the family even if it happens to be an identical twin. Think about how much information is packed efficiently in the fertilized egg to make this happen. And think about how the genes have the ability to interact with the environment to generate more diversity than is present in the genes alone. For example, identical twins are not identical—those closest to them can tell them apart in appearance and in behavior—because of the individually unique interactions between the genes and the environment. In addition, the genetic development instructions are the same in humans around the world but each person born is unique. How cool is that? Personally I'm amazed at the power of the information present in each fertilized egg and the ability of genes to modify their effects continuously depending on cues from the external environment.

We can take this idea farther yet and gain a bit of understanding of how nature "knows" how to pack so much information into a tiny egg. The story goes back to the origins of multicellularity that occurred about ¾ of the way through Earth's history, already briefly mentioned in the discussion of time in Chapter 1. As multicellularity in animals originated, many different innovations in local situations made at least a temporary go of it. Today we can see evidence of this in fossil faunas as well as in a comparison of the developmental patterns of living animals. In the fossil record, most of the organisms have body patterns that resemble our own, at least in the early embryological details. There is an exception in some members of the Ediacarian Fauna of Australia. Included in this fauna are various blobs, discs, and fans that apparently have no relationship to later forms of early multicellular animals. These strange creatures apparently were embryological "outies" rather than "innies" like ourselves.

In our embryological development and in the early development of most other living multicellular animals, after the fertilized egg divides into a ball of cells and then forms a hollow cavity inside, outer cells migrate inside of the developing embryo to form a tube that becomes the digestive tract. Similar patterns are involved in the formation of our lungs and many other features of our bodies. We are "innies". The blobs, discs, and fans of the Ediacarian Fauna were "outies"—their early embryological program caused the migration of their embryological cells outward into evaginations rather than inward to invaginations of their body. If we were "outies", the lining of our stomach and intestine would be surfaces subject to drying and injury outside of our bodies, a situation clearly not as functional as our current "innie" organization. Natural selection obviously took the outies out of the picture.

This harsh filter has been utilized on each detail of human and animal and plant development. Each step of the way, the harsh filter of natural selection functioned within the context of what existed at each time and place. This means that the embryological blueprint used by nature to develop each new human or each new mouse has a long history of utilizing what is at hand and selecting the best available option at each point in time. I can't overemphasize the harshness of the historical process involved.

In each species, humans included, many more eggs, sperm, zygotes, and embryos are produced than can possibly survive. Defects in the embryological program are not tolerated because the narrow conditions that allow life to exist are exacting. People born with "birth defects" actually have very minor deviations from this embryological program I'm talking about—major defects end up as miscarriages or still births. This concept of the narrow conditions in which life can exist was first brought to my attention by a concept paper called *The Knife Edge of Design* in the prestigious journal *Nature* authored by Jack Cohen[41].

Early biologists spent a lot of time looking at how nature functions so they realized only about two offspring per mating pair of animals survived to reproduce in the next generation. This meant that most of the hundreds of eggs in each frog egg mass die to feed the food web. Fortunately, most mice born die before they can reproduce too. I say fortunately because

---

41 Cohen, Jack. 2001. "The Knife Edge of Design", Nature 41:529.

a mother mouse is a mouse-generating machine; they can be pregnant for 20 offspring, nursing 20, and weaning 20 all at the same time! In addition, any defective eggs, sperm, or fertilized eggs die before they even get to start a life outside of the egg. The bottom line is in nature most reproduction goes to feed the food web rather than to become the next generation of parents. The crux of Jack Cohen's concept was the idea that those of us who are alive have passed through a gauntlet of life-threatening challenges to make it into the privileged position of being alive. If anything went seriously wrong during our embryology, we wouldn't be here. In a nutshell, life is perched on "the knife edge of design", a design developed through millions of generations of harsh selection. No wonder there is so much information packed in each tiny fertilized egg!

The embryological development program for each species is ancient and has been built upon the few successful parents of each ancestral species. This means our embryos as well as those of other species go through steps of development that would be unnecessary if the design of the program were done just to make a human or just to make a mouse. Nevertheless, think about how effective nature's system to manufacture new body parts or new individuals is.

In a few weeks in the spring, dormant buds on deciduous trees break open and a genetic development program within the cells of the bud interact quickly with one another to produce fully formed leaves. The body of a clam generates a tough shell to protect the clam's soft body and trees generate bark to protect the living tissue that feeds the leaves of the tree as they grow. Wouldn't it be wonderful if we humans were able to perform this type of information-rich manufacturing for the goods we want and need? All of nature's manufacturing takes place at biological temperatures in the presence of life. Certainly that can't be said of human manufacturing systems. As we work toward a time, species, global perspective and seek to retake our place in nature, it's vital we start looking carefully at how nature succeeds in solving all sorts of challenges. If we learn to take a page from nature's solutions, we will be learning to use information-rich solutions rather than ham-fisted solutions that endanger our existence. Following is one example that is less complicated than making whole humans or mice, but illustrates how specific solutions to specific problems can help us out of many of the evolving problems alluded to in Chapter 3.

Here I'm thinking about poisons to ward off something that is creating problems—like fleas in my house. If nature were designing Orkin's tools, then Orkin would be using a short-lived toxin that affected only the fleas in my house rather than a broad-band toxin that killed all the insects and spiders in my house for several years' duration. For example, in California, there is a species of salamander so outlandishly poisonous that a tiny drop of its poison can kill tens of people. This salamander doesn't bite and you can pick it up and handle it without danger as long as you don't lick your finger or the salamander. It doesn't create any problems for the species around it except those that prey on it. Immediately this story makes me ask why that salamander is so poisonous.

It turns out that these salamanders are eaten mainly by garter snakes—a species of the little green, striped snakes that occur all over the U.S. The garter snakes that live within the range of the salamanders have evolved a resistance to the poison produced by the salamanders, so the salamanders are under positive selective pressure to become ever more poisonous. The garter snakes pay a price for eating the poisonous food, but it just slows them down for a while—those killed by the poison are no longer in the snake population. The moral of the story is that if we're clever, we can find methods to control problems like fleas without using toxins that kill everything around. If we can get our minds around the subtle, elegant, information-rich solutions of nature, we can keep ourselves comfortable without brute force. In short, we must learn to substitute information and cleverness for our current sledge hammers.

Design Rules

The bottom line here is that we must develop a new set of design rules for our living and manufacturing activities if we are going to live peacefully as a part of nature. The four principles outlined above—waste equals food, run on current solar income, depend on diversity, and use information-rich solutions—form a solid foundation. These core principles have led to the complexity, fertility, and wealth now encompassed on earth. They also led to our existence as an intelligent species that has made tremendous strides in understanding this place. If we put our minds to it, we can promote health and abundance rather than resource depletion and death. It's time for us to put away the technologies of the childhood and adolescence of our civilization and to move purposefully toward activities that mirror a mature, stable ecosystem. We have a start with the scientific knowledge necessary to make this happen. Now it's time for us to make the moral step forward into a time, species, global perspective that will give us the spiritual, political, and individual will to make this giant step forward for our civilization. We know a lot about how our home works—it's time for us to begin to live peacefully in it.

Important Concepts of this Chapter

To prosper in the long term, humans must live by nature's operational rules.

Nature's operational rules have developed through earth's history and billions of years of rigorous natural selection so they are well adapted to this place.

In nature, all waste becomes the food of another organism or process. Carbon, nitrogen, phosphorus, and water all recycle continuously in biogeochemcal cycles. Rocks recycle. Food in ecosystems cycles from one organism to another forming food chains and food webs that occur in the world of microorganisms as well as in our world.

On earth, nature runs on current solar income. Energy doesn't cycle. It can be transformed and heat can be transferred from one material to another but energy can't be "made" on earth. Our culture has been "making" energy by releasing the energy stored in the ancient products of photosynthesis.

Nature depends on diversity and mass-production, clones, and globalization of single solutions are antithetical to nature because diversity is the raw material that allows natural selection to choose what is best for each place on earth.

The living world uses information-rich methods of manufacturing and problem solving. These information-rich methods are the products of billions of years of harsh natural selection.

It is our ethical responsibility to redesign our design rules so they fit within the reality of earth's operational rules so we can prosper peacefully in this place.

* * *

So now the background is all in place—we're aware of the time, species, global implications of our choices and we're also aware of how our ancient home works. We also know we are an integral part of nature trying to live healthy lives and raise healthy children while we are trapped in the bubble of unreality that is current American culture. The next step in our common journey is to recognize the challenges we face as we seek to become agents of change in mainstream American culture. This entails taking a deep look at the challenges we face that result from our evolutionary and cultural history as well as those that come from the constraints of the reality around us. We can't envision a different future unless we fully appreciate why things are as they are.

Section III

The Challenges We Face

Chapter 6

The Commons Concept

Every important resource humans and other species depend upon is vulnerable to exploitation, not because people or other species are evil, but because that's how this universe and our existence is organized. The most foundational resources on earth today are air, water, soil, our social order, climate predictability, biodiversity (the heritage of genetic diversity present on the globe), space to live, and a means to make a living. Easter Island serves as an example of the general principles involved in vital resources held in common with other individuals.

Easter Island

Rapa Nui, also called Easter Island, is the world's most isolated piece of land, an area of only 64 square miles, located 2,000 miles west of South America and 1,400 miles from the nearest habitable island. Easter Island is famous for its giant stone statues; some of them even have big red stone crowns. It is also famous because when it was discovered by a Dutch explorer in 1722, it had only about 2,000 desperately poor people living in a grass wasteland with no native animals larger than insects and no trees taller than 3 feet. The fame comes from the mystery of how these people or some other long-gone inhabitants had been able to make and erect the statues. The island's history has been reconstructed by the combined work of archaeologists, fossil pollen experts, and paleontologists who identify species from their bones and/or shells. Different groups have reported different interpretations of the data,[42] but all agree that the island was colonized by Polynesians sometime between 800-1200 AD. The Polynesians arrived in boats made of hollowed-out trees stabilized by out riggers and loaded with seeds and animals to start their food system—chickens, bananas, taro, sweet potatoes, sugarcane, and paper mulberry[43]. Polynesian rats also arrived with these first colonists[44]—no one knows if it was on purpose or not.

The early colonists found a tropical island paradise that met their every need. The giant palm trees and other native trees could be made into more boats, fiber for ropes, firewood, and wine for drinking. With their big seaworthy canoes, they hunted for porpoise and other sea food far from the island. They also feasted on shorebirds and their eggs, seals, and on the Polynesian rats, at least until all of those were extinct. With plentiful food and time on their hands, they set up a thriving economy in quarrying stone, making the giant stone statues (900+ of them each weighing up to 90 tons), and transporting the statues from the quarry to near-shore platforms. Transportation would have been done by rolling the statues along on the trunks of the giant palms. By the 1600s, wood became scarce, and the trees and shrubs were gone by 1722 when the Dutch rediscovered the island.

While the ecology was still intact, the population grew, different tribes apparently developed, and special interest groups

---

[42] Jared Diamond. "Easter Island Revisited" Science, 21 September 2007, vol 317, pgs 1602-4.

[43] Jared Diamond. "Easter's End," Discover, August 1995, pgs 62-69.

[44] Terry L. Hunt. "Rethinking the Fall of Easter Island," American Scientist, Vol 94, pgs 412-419.

evolved with roles like farmer, stone cutter, tree cutter, statue transporter, and so on. The people living there must have seen how vulnerable their way of life was to the loss of their forests—without the giant trees, they couldn't make seaworthy canoes and without the canoes, they could no longer harvest food from the sea except that found near shore, most of which was extinct already anyway by the time the forests were in decline because the marine shore life had already been exploited to extinction as a source of easily available food. Nevertheless, they kept on course, possibly helped along in the deforestation by the rats eating palm nuts. As the ecology collapsed, so did their civilization.

The ensuing devastation would be worthy of situations like those depicted in movies like *Road Warrior*, *The Postman*, or *Waterworld*, but worse. They broke into warring factions that raided one another for food. There is evidence of cannibalism. Can you imagine your child in your rival's cooking pot? Many lived as refugees in caves. Finally, even this system broke down so all that remained by 1722 was a small number of desperately poor people with limited fuel to cook their meager supply of food. It is possible that the European practice of taking people for slaves may have exacerbated the problems. Regardless, the fate of Rapa Nui is an example of the catastrophic future we face if we allow our economically driven decisions to drive our civilization to devastate our environment. It's important we recognize the role our "giant palms" play in our well-being and learn the lesson of Easter Island.

We can ask why the Easter Islanders didn't conserve their forests, especially the giant palms, because those were so important to their soil quality and the ability of other species of trees to live. We can also ask why they didn't protect the populations of wildlife on their island that provided such nutritious food. Of course, we can only speculate on the reasons, but I suspect if there were some environmentalists there, they would have been only a few voices among many—most would have been in favor of continuing to cut the trees faster than they would grow because their livelihoods (i.e., their economy) depended upon the uses of the trees. And if they banded together to stop the cutting, almost surely there would have been a few cheaters who would sneak out at night and cut the biggest tree left to make themselves a new seaworthy canoe. In addition, we humans are amazingly resourceful—when one resource becomes scarce we shift quickly to other resources, so probably when the last palm was cut it had very little economic importance. The last factor contributing to the destruction of society on Easter Island was probably how our minds are adapted to change. We react quickly when sudden changes occur, but it is easy to miss changes that occur over one or more generations. Slow change just doesn't get our attention—slow change becomes part of "normal," just as we've accepted the changes associated with the industrial revolution that have dramatically altered the earth's life support systems in less than 200 years. So succeeding generations wouldn't have been aware of the gradual loss of the forests. The driving forces at work on Easter Island and at work in our globalized world today can't be fixed easily.

The Tragedy of the Commons

Back in 1968, Garrett Hardin published a landmark paper in the journal Science titled *The Tragedy of the Commons*.[45]

---

45 Garrett Hardin, "The Tragedy of the Commons", Science, Vol. 162, December 13, 1968, pgs 1243-1248.

The paper is targeted at human population growth, although the commons concept has a much wider application to our everyday lives and to life on this planet. Hardin distinguished problems according to whether they have a technical solution or not. Technical-solution problems are those that can be solved by developing new technologies that bypass the drivers creating the problems; these kinds of problems don't require changes in human values or ideas of morality. Many of our modern environmental problems have no technical solution and therefore are systems that create a tragedy of the commons unless we change human morality and behavior.

Hardin explains the tragedy of the commons as follows. In the old English countryside, peasants had individual houses and livestock but shared a common pasture. As long as there weren't too many people or animals, which was the normal case until social stability was attained, there wasn't a problem because there was plenty of pasture capacity for all. But with social stability came an increase in population. Individual peasants would benefit by adding another cow to their herd. By doing this the peasant got all of the benefits, but the damage caused by increased grazing was shared by all peasants who used the commons. Each peasant in the community would be in the same boat—adding another animal to the pasture worked to their benefit but harmed the shared resource of the community. Essentially in a commons, *the profits are privatized and the costs are socialized.* Because of the unbalanced benefit-cost equation, the inevitable outcome is an overgrazed common pasture unable to support any cows.

Commons in Our Lives

The tragedy of the commons has a profound impact on our lives because we live in and are surrounded by commons of all sorts. The key characteristic of a commons is that everyone has a right to use it without paying but no one owns it or has exclusive rights to it. Some important commons are air, water, soil, national parks, darkness, oceans, biodiversity, Antarctica, the Arctic, global climate, ecosystem services like production of O_2, quiet, public roads, the Internet, broadcast bands on the airwaves, government tax money, the stability of the economy, darkness at night, and so on. Some of these may not seem obvious because we are so used to taking them for granted.

For example, we take odor-free air as a given until some industry moves next to us, like the corporate pig farms in Oklahoma, and we find we can't live in our homes without a gas mask. Odor-free air is trespassed on daily—have you never been in a public place and been assaulted by second-hand smoke or some overpowering perfume? Darkness is a commons we under appreciate until a close neighbor installs a spot light that trespasses on our bedroom window. The impacts can range from sleeplessness to an increased incidence of breast cancer in women because the light at night interferes with key hormones from the pituitary gland. Quiet is another non-obvious commons easily taken for granted but frequently trespassed upon—it's hard to notice the benefits of quiet (or choosing your own noise) until they're gone. David Quammen, in a wonderful book of essays called *The Flight of the Iguana*, comments that if you put a Cocker Spaniel in a fenced yard where it's bored, you have a created a barking machine that will never stop. Traffic noise falls in this category, especially loud bass music and hot trucks and motorcycles with fancy mufflers—these are mobile assaults on the peace and quiet of every species living near roadways. Paris now has strict noise pollution laws to protect the commons of quiet because one loud motorcycle can wake up 300,000 people as it travels through the city at night.

The commons mentioned in the last paragraph are nuisances that can create health problems but there are other commons that underpin our very existence: breathable air, oxygen, clean fresh water, fertile soil, nature's ability to maintain the cycling of plant nutrients and atmospheric components, nature's ability to continue adapting to change through evolution, and space for each species (including humans) to make a living and nurture their offspring. Possibly you can think of other commons that are foundational to life.

For each commons, there is a dilemma because, as Hardin points out, you cannot appeal to the conscience of people to take care of the problem. In other words, educating people will not solve the problems of the commons because some people have no social conscience, which is the key factor that must be activated to instill a sense of fair play. Essentially the world is divided into what I call slobs and suckers. The suckers are those who pay attention, learn, and then modify their behavior to help the situation. The slobs do what they want or what profits them, and they are happy they don't have a lot of competition when the suckers are careful of their own actions. I suspect the Cocker Spaniel owner above would not be happy with neighbors who moved in with a Great Dane who could "out-bark" their Cocker Spaniel and the light-loving neighbors would be up-in-arms if it were their bedroom window lighted up like day, yet neither would be aware of how they have damaged their neighbors with their own trespasses.

Each of us is a slob or a sucker in many respects. For example, I am a slob by virtue of living in a house in the country that has conventional plumbing. My country location is important because each time I drive to the city, I am polluting the air with my auto exhaust and creating noise for those who live closer to the city than I do. Conventional plumbing makes me a slob because each time I flush a toilet or use some hot water, excess water is used from our common aquifer and/or energy is wasted. Similarly, I'm a sucker when I manage my land as toxin-free as I can in order to keep the creek behind my house as clean as possible. Each neighbor in the entire watershed who uses lawn chemicals is a slob with respect to the cleanliness of the water that flows into Lake Michigan. Because each of us is so embedded in the tragedy of the commons and because our species has become so ubiquitous, we must use our intelligence to control the problem.

Social responsibility is the product of definite social arrangements, not the product of propaganda (aka education). The closer we live together as the human population grows and becomes more dense, the more "definite social arrangements" we must have if this planet is to remain livable. This dilemma is written big and small all around us. Humans aren't the only ones who will trespass on a commons.

Woodchuck Slobs

A few years ago, I unwittingly set up a commons in my back yard—not a commons among people but rather a commons among me and my animal neighbors. Because I have so many animals in my house, I also have a lot of organic waste—partially chewed apple sections, uneaten greens when the lizards don't clean their plate, squash pulp and seeds, and the list goes on. As an environmentalist who is trying to lighten my footprint on this small piece of land, I figured the best thing to do with all the organic waste from the kitchen, lizard litter box (yes, they are potty trained!), garden, pond, and lawn was to make compost to spread on the gardens. I had developed a system down near the creek with two structures made out of waste wood—one for the fresh stuff and the other with chipper-shredded material that I let compost for a year or so. The compost worms I had ordered off the Internet were living happily in my compost piles, making rich soil from the wastes.

Things were going pretty well except I found my compost didn't have much "zing" after my snake died of old age. The snake was important to the quality of the compost because the snake ate mice and the mice bedding added animal urine and feces to the mix. Of course, after the snake died, I got rid of my mouse colony. Without the dirty mouse bedding, I was short of nitrogen—because of this experience I eat meat from locally-grown animals whose feces is used as compost to fertilize fields. Anyway, back to the compost pile.

As things went on I started having a harder and harder time growing a garden. Shoots were chewed off just about as soon as they sprouted and bedding plants were whapped off in big swathes. For two years I hurried around in the spring getting the garden ready for the summer and then left it in the care of a student while I went to Madagascar doing research for a month each time. When I got back, I was always anxious to see how big the garden was only to be disappointed by what can only be described as a disaster. I figured the student was just doing a poor job, but that wasn't the case.

Last summer I was home and discovered I had unwittingly set up perfect conditions for woodchucks. Oh yes, woodchucks moved in just as the Polynesians had moved into Rapa Nui, and they were doing really well until I started paying attention to the situation. They had dug dens all along the stream bank and up under my compost pile. Apparently I had been feeding them all winter with my daily kitchen waste as well as all summer with my garden. The upshot of this story is that I personally know of six woodchucks that either died on the road, died in my kill trap, were captured in a live trap and relocated, or died in the inlet to my pond. Each summer is the summer of something—last summer was the summer of the woodchuck. Here's the story and it's mostly a sad one.

My first course of action was to live trap and relocate the woodchucks; I looked on the internet to find out what could be used as bait in a live trap for woodchucks—not trivial because they seemed to love the garden so much. Nothing seemed to work—they wouldn't take any bait I offered—fresh corn, peas, green beans, nothing. Meanwhile the garden suffered horribly and I was getting more frustrated by the day. I sneaked around the backyard, crawling along on my belly trying to catch them in the act. No luck for the longest time. The first woodchuck I saw was a big one I found dead out on the road. That's sad, but frankly I was happy about it. I thought my woodchuck problem was solved. Not so. The next batch of kale and collard bedding plants was again promptly eaten down to nubbins. I set a Duke #220 kill trap right around one of the woodchuck holes. In short order, I had trapped and killed another big woodchuck. That was the good news. It was also bad news because the dead woodchuck was a female who was obviously nursing a batch of young. She had eight teats, an observation that brought fear into my heart. I was hopeful the young were too immature to make it on their own but felt bad about them being starving orphans who had lost both parents. Still, eight of them!

Things were quiet for a couple of weeks, the garden was again growing, and I felt I could raise enough food for myself and my animals. Then the woodchucks started striking again. By now I had gotten so I could spot them and began to understand their daily cycles. They would show up about 10 in the morning and again in the late afternoon. These woodchucks were young so I tried the live trap idea again. Luckily I caught two of them over a few days and was able to move them to good woodchuck habitat that wasn't near anyone's garden. They are actually kind of cute—about a foot or more long, with a rolly body and short legs. They are a type of rodent so they have big incisors and little round ears. But they are also garden vacuums so they had to move out of my neighborhood. I kept the live trap out and freshly baited but didn't catch any more woodchucks. A neighbor across the way was having a horrible time with his garden, too, and his solution was to shoot the animals that came his way. For a while I felt good that at least a couple of this family had made it in a spot where they weren't devastating someone's garden.

Then one day I was walking down to my garden and surprised another woodchuck. To understand this, you must understand how my garden is set up—not a typical garden but it produces enough food to take care of most of our vegetable needs for the winter because I preserve food by canning and freezing it. Anyway, I garden in planting boxes built into a hill that is full of freshwater springs. When the hole for my basement was dug 30+ years ago, it kept filling with spring water. The builders had the good sense to collect the water into a drain tile and channel it down the hill. My son, Darren, and I used that water source to build a 30' diameter pond, and then we built planting boxes up the hill around the pond. This works really well because the planting boxes are naturally irrigated by the spring water that seeps out. We built the planting boxes right over the drain tile from the basement and set it up so the pond is fed by the house's spring water and overflows through a grassy waterway into Little Bass Creek, which runs behind my house. In this wonderful setting I raise everything from water lilies and lotus to potatoes and tomatoes. And woodchucks.

The little fellow I surprised was on the Tennessee flagstone next to the pond—if he had turned south toward the creek, he would have gotten away with no problems but instead he turned toward me and scurried up the drain tile that feeds water into the pond. Oh, what a fatal mistake. I figured the woodchuck would wait until the coast was clear and then back out and go on its way so I sat quietly waiting. After about five minutes, I heard a lot of thumping and scratching and then no more noise. I figured it had dug its way out of the drain tile, which would have made a real mess for me because the tile was buried under tons of dirt and railroad ties. To repair the damage would have cost untold time and expense. I don't know what you would do, but I decided to give it some time to see if I really had to dig up the whole yard.

Within a couple of days the water flowing into the pond had dropped to a tiny trickle, an observation that raised another possibility—the woodchuck might have died, bloated up, and plugged the tile. Either that or water was finding a new channel through the woodchuck's new hole in the tile. Bad news either way. The plumber/sewer company I called said it would cost more than $300 to clear the tile.

By now I was pretty sure the woodchuck—I called him Plug—was in the drain tile and dead because the water that did come out created smelly green foam. Both of my kids are engineers who gave me good suggestions on how to deal with the problem—Darren suggested a variety of intricate methods to extract Plug, and Adrienne thought I should just be patient and let it rot. Both seemed to be good ideas and I tried them alternately. On one of my let's-get-this-solved-now days, I tied fish hooks onto my garden hose to snag Plug and pull him out—by the time I got through trying that he was a good 20' up the tube because he slipped forward easily, but his arms acted like brakes when I tried to pull him out. So I alternated between new plots to get the woodchuck out and being patient waiting for it to decompose. After several weeks, water started coming out all around the planting boxes and the grass between the house and garden was growing as if it had been watered and fertilized really well. Finally I figured if this kept up, I might end up with a flooded basement.

With no options left, I called the plumber and in short order two fellows with big boots and gloves came out with a super-sized rotor-rooter. About 20' up the drain tile, they snagged Plug and pulled him out. As he came out, both men jumped back and said, "Ewww!" because by this time Plug had lost all his hair and stank a bit—but not much, because apparently the cold water of the spring had helped to preserve his body. The rotor rooter had ripped a hole in his stomach, which was full of bright green, well-preserved kale, green beans, and collard greens. I gathered him up and tossed his body out in the ravine for the carnivores to eat, paid the price, and was happy to see that the water immediately began flowing into the pond again. These guys offered to install a woodchuck guard over the tile but I did it myself by putting chicken wire over it. You'd think that would be the end of the story, but there's one more piece before I had peace.

A few days later, another woodchuck showed up. As usual for all but the youngest woodchucks, it wouldn't go into my live trap, so I again set the Duke #220 trap over a woodchuck hole. In a few days, it was dead. Hopefully, that's the last of that family of woodchucks. There are a few lessons in this story.

One harkens back to the cheese—these woodchucks set up housekeeping at a very stale cheese station and stayed there even with the danger of the road and of my traps. They also set themselves up to make a living off of me and my efforts. Neither is a wise choice. From another perspective, woodchucks have to live somewhere and the system I had set up was a perfectly reasonable spot for them to make into a home for raising their babies. I am not without blame in the tragedy either. Obviously animals will move in if you feed them a reliable 3-course meal each day. And the best way to keep pests out of your garden is to move it every couple of years or, at a minimum, fence it in so things like woodchucks can't enter. Believe me, I thought about both, but I don't have that much space and it's impossible to fence the edge of a creek. I could also have quit gardening for a few years and the woodchucks might have moved on. There's a lot of blame to go around, but the key message is this patch of land that is my garden is a commons. Both I and the woodchucks felt we had a perfect right to the earth's bounty in that spot, and because the garden plants couldn't support us both conflict arose. Not a pretty situation. We humans are engaged in major conflicts and headed toward larger ones as our commons are depleted and damaged, much bloodier and worse than the conflict in my back yard.

The Tragedy of the Oceans

On a global scale, the oceans may be the commons in most imminent danger of catastrophic collapse. As seafood has become an everyday staple of Americans coupled with wide adoption of our industrialized, mass-production technologies, the unthinkable has happened— oceans have hit the limit of their ability to rebound from abuse. Most of us are slobs in our relationship to the 70% of the earth that is covered with saltwater. Champions of the seas are few and far between and their voices are a mere whisper compared to the as yet unfettered greed of most of us. The total collapse of this commons, i.e., its inability to support *any* human nutrition, is a short 40 years away[46]. The oceans now are comparable to Hardin's common pasture when so many cows are grazing that the plants are cropped off before they can grow.

In some sense the sea is our mother—early humans spread from Africa to other lands along its edges and still today about 1/6 of humans rely primarily on the oceans for food; many more rely on it for their living. Even though we live mainly on land, our lives are intricately underpinned by the free services we receive from the oceans. These huge bodies of water are critical to the cycling of carbon in the biosphere and the generation of oxygen that enters the atmosphere from photosynthesis done by the meadows of tiny plants at the ocean surface. Ocean currents warm or cool various landmasses. The oceans are a huge heat sink—without them global warming would be making much more profound impacts on climate. Biodiversity at the ocean edges is important for flood control and its species are important in waste detoxification of runoff from our cities. Banyon trees

---

46 Boris Worm, Edward B. Barbier, Nicola Beaumont, J. Emmett Duffy, Carl Folke, Benjamin S. Halpern, Jeremy B.C. Jackson, Heike K. Lotze, Fiorenze Micheli, Stephen R. Palumbi, Enric Sala, Kimberley A. Selkoe, John J. Stachowicz, and Reg Watson. 2006. "Impacts of Biodiversity Loss on Ocean ecosystem Services," Science vol 314, 3 November 2006, 787-790.

at ocean edges mitigate terrestrial damage from catastrophes like tsunamis. I could go on. It would sound like one of those eulogies for great and famous people because the oceans do so many things for us. The death of the oceans is not something that can occur without severe repercussions—many more mass deaths like the Indonesian tsunami death toll, but also many quiet deaths from lack of food.

The loss of the oceans' bounty is something humans can hardly comprehend. Years ago Charlton Heston starred as a New York detective in *Soylent Green*. This 1973 movie was about a futuristic earth that was severely overpopulated; crowded masses of people were subsisting on a manufactured foodstuff called soylent green. Supposedly soylent green was made from the meadows of tiny ocean plants called phytoplankton. They were eating phytoplankton because their land was so damaged that it was no longer fertile. Heston was marked for death because he had discovered that the oceans were dying, and to provide food the industrial food system was using people's bodies to make soylent green. Life was so horrible that people were willing to die in exchange for a few minutes in a private room with a view of meadows and trees. This was a horror story that got it backward—their land died before the ocean—we assume this because we are so confident that the oceans are too huge for us to ever deplete their resources.

Industrialized Fishing Technology

I suspect if exploitation of the ocean were limited to one person, one fish hook, the nearly 6.8 billion people of the earth couldn't fish all the fish out of the sea. But that's not the case. Just as we have industrialized food production on land, we have also industrialized fishing. Now huge ships called factory trawlers move over the near shore ocean areas dragging nets so big that you can fit twelve 747 jet planes in the net opening. These nets rip up the ocean floor, an ecosystem comparable to the old growth forests on land. They catch every swimming thing in their path regardless of whether they have a license to catch that species or not. Ships are licensed to catch only certain species so crab and other edible species are wasted as by-catch. At least half of what is caught and killed is thrown back into the sea as by-catch. The other half is processed on board the ship into the frozen fish products sold for pennies on the market.

Other industrialized technologies are used for the open ocean. Schools of fish are located by using satellite images and GPS technology and then they are fished using drift nets that are miles long. The catch includes sea turtles, dolphins, seals, and anything else that happens to swim in their path. Longlining is another technique commonly used in the open ocean. Miles of huge baited hooks are dragged behind ships that catch and kill sea birds, turtles, sharks, and many other species, all considered by-catch. With these techniques and with the increased pressure of sports fishing from wealthy countries and subsistence fishing from poor countries, the food webs of the ocean are crashing.

Killing the Food Web

A bit of biological background is important here to understand the severity of the situation. As discussed in Chapter 5, ecosystems that have evolved to a stable, sustainable system are made up of communities of organisms that form a food/energy pyramid. The base of the pyramid is the plants that capture the sun's energy and transform it into energy in the form of the chemical bonds of the molecules that make up their bodies. The next layer upward is the herbivores, animals that eat plants as their source of food energy. Above them and fewer in number are carnivores that eat the bodies of the herbivores. There can be more layers upward, each smaller than the one below—the rarest species in an ecosystem will be those carnivores that eat carnivorous carnivores. Lobsters are a great example of a top carnivore. Of course, few of these species eat only one other species so this whole network of eaters and the eaten forms a food web. The more ancient and diverse the ecosystem, the more intricate and complicated the food web. The more intricate and complicated the food web, the more stable the ecosystem.

Our abuse of the oceans has led to fishing down the food web and tearing up the base of the food/energy pyramid. A few years ago we were eating swordfish and mackerel. Then we decided sharks were good to eat and now we're eating the fish they used to eat. DNA testing on fish from the supermarket labeled "pollack" or "haddock" reveals that the fish is usually not that species at all. Because these huge fisheries operations must make a profit or go out of business, they bring home and sell anything they can pass off as a legal fish and concurrently leave a swath of ocean death in their path.

One regional fishery after another has collapsed since 1970 and there seems to be no end in sight. Each fishery collapse is a step in the loss of biodiversity in the oceans and each piece of lost diversity reduces the stability, recovery ability, and water quality of the oceans. Global deep sea extinctions are slowly being uncovered but it is well known that estuaries, coral reefs, and coastal oceanic communities are losing populations, species, and entire food webs[47]. Because the oceans are so huge and complex, it's almost impossible to have accurate data on what's

---

47 Worm et al. 2006. Cited above in footnote 46.

going on, but according to *National Geographic*[48], nearly a third of the world's fish are clearly overexploited. The damage is the worst in the Atlantic; there and around the world much of the fishing is illegal. The rate of fisheries collapse is accelerating, and cumulative collapses include 65% of recorded species[49] —notice we don't know anything about the species lower on the food web that are killed as by-catch.

A reasonable person at this time might ask why no one told them this was going on. The fact is we have been told, but we haven't heeded the warnings. There is still plenty of seafood on the grocery shelves and the price is still reasonable. But if we are willing to hear, the messages are there before us. For example, *Time* magazine ran an article, "Oceans of Nothing" in the November 13, 2006 issue that clearly showed the relationship between diversity and risk as well as the projections of total collapse of fisheries by the year 2048.

The people doing the fishing know what's going on—they are catching fewer and fewer fish per amount of fishing effort. In some areas like the Newfoundland cod fishery that fed people around the world for hundreds of years, industrial fishing has collapsed the entire culture of people who have made their living for centuries using conventional fishing practices[50]. The slob and sucker problem is written large in the cod fishery collapse. No amount of education or individual self-denial would have kept this commons healthy once industrial fishing arrived on the scene.

---

48 Fen Montaigne. 2007. "Still Waters," Special Report, National Geographic, April 2007, pgs 42-69.

49 Worm et al. 2006. Cited above in footnote 46.

50 Chris Carroll. 2007. "End of the Line." Special Report. National Geographic. April 2007. Pgs 90-99.

This is starting to sound like a dirge—no hope and no way out of the problem. But that's not true. The solutions are clear and we've known them for quite some time. The only problem is that we people—lots of us around the globe—must demand definite social arrangements in our own country's waters and in the open ocean to bring about ethical and behavioral changes in humans' relationship to the ocean. First let's talk about some things that are *not* solutions.

Non-solutions

The highest thing on my list of non-solutions is to allow the current trend in increased fish farming at ocean edges and in the open ocean to escalate in response to decreased yields of wild-caught fish. Fish farming in the oceans and along the shores makes the ocean ecosystem collapse problem invisible to consumers. It also is speeding the destruction of the ocean's fecundity. Fish farming is occurring in the form of blue fin tuna ranchers who fatten Mediterranean tuna before slaughter by caging them and supplementing their food, logic similar to what is done on cattle feedlots. Ocean farming is also occurring in deep-water cages supported by pillars; the fish are fed grains and fish meal[51], a practice that puts further stress on the ocean food web and raises the possibility of introducing terrestrial diseases into the ocean ecosystem. Near shore areas are farmed heavily for shrimp, tilapia, and other popular species. To make way for this farming, all over the tropics the shoreline giant banyan trees are destroyed. Banyans have evolved as the keystone species of a biodiversity haven. These trees make shorelines abundant with sea food and they protect shorelines during hurricanes and

---

51 Carrol, 2007. Cited in footnote 50.

tsunamis because their multiple roots form a near-shore, above-water forest. These trees and their companion communities, which are full of many species of edible fish, are being cleared to make way for near-shore monocultures of shrimp and talapia farms. This is bad technology because it violates two of nature's operational principles: nature depends on diversity and waste equals food. Havens of biodiversity that recycle nutrients into other food sources are replaced by monocultures that produce massive amounts of organic pollution.

The other practice that should be abandoned is management of the oceans by controlling the catch limit of individual species. This is how we've always done it, but the current ocean crisis proves this method doesn't work. If you think about it, monitoring the catch of a single species one at a time by various controlling agencies could never work because it's like the modern U.S. medical care system—lots of specialists treating one or another aspect of the patient's body while the patient dies from poor health. Individuals simply don't have enough information to make sound decisions or even good guesses with this dispersed system. Further, it puts fishermen and their equipment at risk. If a bad storm occurs on opening day of the season, the fishermen still rush out to catch as many fish as they can—if they wait, there are no fish left for them to catch. Obviously there is also the problem of by-catch. Catch limits are counted on shore and don't take into account the countless individuals accidentally caught, killed, and thrown overboard as by-catch. Fortunately there are solutions available that will ensure the health and vitality of ocean ecosystems into the future.

Successful Social Arrangements

To save the oceans and their myriad ecosystem services as well as their bounty of food, we must immediately set aside a minimum of 10% of coasts and ocean surface as completely off limits to any kind of extractive activities. Ten percent is a minimum—the ocean would actually produce much more food if we set aside somewhere between 20 and 50% of its open water area and coasts. That seems like a lot, but the fact is that we could harvest much more food out of half the ocean than we do now out of the whole ocean if we would keep our hands completely off the protected half.

This solution would work because nature is naturally abundant. Evolution works by each species producing many more offspring than can possibly survive on the limited resources available. This means there are always extra young produced. If half of the ocean and its shores were off limits to fishing, then there would be a place where both long-lived and short-lived species could produce offspring. The density on the reserves would become very high in short order and then the process of spillover would occur—the eggs and young would constantly colonize areas outside of the reserves while the reserves would protect the ability of species to reproduce. This same thing would be true of the entire food web so the biodiversity and resilience of the ocean would be forever protected.

The slobs in the situation are of course fighting having any reserves and are trespassing on the boundaries of reserves that are already in place. And when they lose the argument about having reserves, the slobs want to move them periodically so they have the opportunity to go into the breeding stocks and swoop up the biggest fish and make huge temporary profits. But we don't want to kill the goose that lays the golden egg—the biggest and oldest fish are worth much more as parents to a new generation of fish than their value as increased profit and fish steaks for a few.

This kind of solution may seem impossible but I assure you it isn't. Within five years of the creation of marine reserves in Florida and St. Lucia in the Caribbean, local fish catches increased between 46 and 90%[52]. New Zealand's first marine reserve was opened in 1977[53], a move that was fought hard by most local fisher people. These same people now are highly enthusiastic, and New Zealand already has set aside 31 reserves and plans to have 10% of its coastal water in reserves by the year 2010. This forward-thinking country has reaped many benefits from its foresight—in addition to revived fisheries, people now come from all over the world to see the underwater Gardens of Eden that grow quickly when the heavy burden of human greed is removed.

As Garrett Hardin made clear, we can only protect a commons through definite social arrangements. These arrangements enforce moral behavior (sucker behavior if you will) in the use of the commons. In the years since Hardin's paper appeared, a whole field of study has developed around the tragedy of the commons, a testament to human creativity. One key theme appears over and over in this body of knowledge: the best way to protect a commons is to privatize both benefits and cost. In other words, those who protect the commons are the only ones who can benefit directly from the protection.

---

52 Callum M. Roberts, James A. Bohnsack, Fiona Gell, Julie P. Hawkens, and Renata Goodridge. 2001. "Effects of Marine Reserves on Adjacent Fisheries." Science, 30 November 2001, pgs 1920-1923.

53 Kennedy Warne. 2007. "Blue Haven." Special Report. National Geographic, April 2007, pgs 70-89.

An article in the journal *Science* by Quentin Grafton, *Economics of Overexploitation Revisited*[54], used economic modeling to study the profits and revenues of global fisheries given their current depleted state. Their results may bring peace between the fisheries industry and the conservationists so a way forward can be agreed upon. This type of agreement is critical because there are billions of dollars and economic stability at stake in any decisions made. The solution hinges on the fact that when fish are more abundant it takes less time and fuel to catch the quota allotted, making the whole enterprise more profitable. The key piece necessary to make recovery plans possible, though, is that those who cooperate with ocean recovery plans would be the *only* ones who later get to fish. In other words, fishing corporations and individual fishermen would give up current income and reap higher profits in the future.

More recently, Christopher Costello, Steven Gaines, and John Lynham[55] wrestled with the problem of the commons in ocean fisheries. They studied 11,135 fisheries worldwide and found only a few collapses of fish stocks in fisheries where fish-stakeholders own the rights to a certain percentage of the catch, a system called individual transferable quotas. If the stocks are lower than expected, then all the stakeholders have a reduced share. Each stakeholder in this system benefits more by taking care of the commons than by exploiting it. They take a smaller portion now so they can have larger profits in the future. The

---

54 R.Q. Grafton, T. Kompas, and R.W. Hlborn. 2007. "Economics of Overexploitation Revisited." Science Vol 318, 7 December 2007, pg 1601.

55 Christopher Costello, Steven D. Gaines, and John Lynham. 2008. "Can Catch Shares Prevent Ocean Collapse?" Science, 19 September, 2008, pgs 1678-1681,

key to the whole system is a limited number of shares—people who did not contribute to begin with cannot be allowed to come in later and help themselves to a share of the fish. Professional fishermen are not the only ones who must be restrained.

Private yachtsmen must be tightly controlled. I hadn't realized how intense the yacht fishing pressure was off the coasts of the U.S. until I went to New Hampshire with my sister Mary Jo one summer. We went on a whale watching tour out into the Atlantic. After several hours of steaming away from shore, the ocean was still dotted with huge yachts, almost like a parking lot. Each boat had multiple fishing lines in the water, but I suspect few were catching anything from what I saw.

Their collective activities have a huge impact on fish abundance in the ocean, partly because there are so many private yachts. Possibly the solution with private yachts might be a lottery for fishing permits, similar to goat-hunting permits in Colorado or bear-hunting permits in Michigan. One way or the other, there are solutions to the tragedy of the commons taking place in our oceans. The same sorts of solutions can be used for carbon emissions to the atmosphere and other foundational commons we humans share because we're citizens of this world.

None of these plans will work if slobs aren't completely controlled in each commons. It will take more than a set of rules to make any system work because "Cheaters love rules and regulations. Every time you put up a barrier, they figure out a way around it."[56] In other words, it will take massive local, regional, national and international cooperation along with investment in serious policing at all levels to ensure adherence to ocean or atmospheric protections. It will also take individual

---

56 Tina Bart in the novel "P is for Peril" by Sue Grafton

attention from each of us to recognize and ensure protection of those commons on which we all depend. As in so many other instances, our purchases (see the Environmental Defense Fund web site below for updated lists of ecologically safe seafood purchases)[57], our votes, and our supervision of our elected officials are critical to ensuring ethical behavior at all levels in this place where we live. Ethical behavior requires that we recognize the commons upon which we all depend and actively support local, regional, national, and international actions to protect each commons from damage.

---

57 http://www.edf.org/page.cfm?tagID=1521&redirect=seafood

Important Concepts of this Chapter

Commons, the things we all have a right to but nobody owns, are essential to our survival for the long term. Many problems associated with commons are no-technical-solution problems. They can only be solved by a change in human behavior.

Human behavior can only be changed by a change in the system that enforces new standards of ethical behavior.

Education alone is insufficient to solve the problems of a commons because education appeals to the conscience of people and many people don't have a conscience, especially when they gain all the profit while the damage is shared by everybody.

The commons in most imminent danger of collapse is the oceans, which are predicted to produce no sea food at all by the year 2048 because modern technology has made it possible to exploit ocean resources far past their ability to regenerate.

Definite social arrangements can stop the tragedy of the commons occurring in the oceans by using ocean and near-shore reserves that are off-limits to any extractive activities. These areas seed the rest of the ocean and shores with species that are used as sea food. The rest of the ocean and shores can be protected by transferable quotas that ensure fishermen have a stake in protecting the ocean commons.

It is critical that we use our individual power to demand definite social arrangements to protect all of the commons upon which we depend.

* * *

When we drive our decisions mainly by economic concerns, humans and the environment pretty consistently lose. The next chapter focuses on food and water, issues unarguably crucial to human health and welfare, and how the I, me, now perspective coupled with apparently cheap choices have cost us more than we can possibly afford.

Chapter 7

Food and Water

With our old industrial model of mass-produced, pollution-generating, resource-depleting technologies, the danger of making consumer choices based primarily on economic factors is acute. In the long term, we must protect the commons of our soil, water, and air in order for humans and other species to live on this planet. Fundamental among our needs are food and water, things too important to trust to profit-driven corporations using the old industrial model that has been adapted to food production. The following story illustrates the primacy of economic concerns in the affairs of humans and the inevitable tragedy of the commons that results.

The War with the Newts

Karel Čapek, a Czechoslovakian writer from the early 1900s, published a wonderful story called *The War with the Newts*[58] that was later translated into English by Ewald Osers. Here I'm giving you the SparkNotes version because the story gives such insight into the central role economy-first decisions play in our civilization's rush to catastrophe.

---

58 Karel Čapek, "War With The Newts," original title "Válka s mloky," 1936, Translated from the original by Ewald Osers, UNESCO Collection of Representative Works, ISBN 92-3-103599-1.

A ship's captain discovers a new species called newts on a remote tropical island. Newts walk upright, are taciturn, and live in colonies in shallow lagoons. Their main predators are sharks against which they have no defenses other than leaving their homes and coming out onto land. In the first contact with humans some newts are killed and later dissected, others are taken captive and returned to the homeland with the kind of fanfare typical of human endeavors—parties, pictures, and pageantry. Of course, as is typical of humans, the first thought is how to make money from these creatures—how to use them for some commercial purpose.

In short order, the newts are being sold and transplanted into shallow harbors around the globe with the purpose of keeping the harbors dredged of the sediment that keeps ships from safely coming to shore. All the newts get in return is metal that they use to protect themselves from sharks and explosives they use to make their harbor-clearing work quicker and easier. Meanwhile the scientists do the typical suite of studies such as crushing the inner ear of captive newts to determine the effect of this mutilation on their behavior. Newts are able to quickly learn to interact with humans using proper English so they are also used as servants.

The system evolves, as all systems do, and soon the newts have been spread around the globe and whole sectors of the global economy are based on the trade with the newts and reliance on their activities. Then one day a residential area near a harbor suddenly collapses into the ocean. Apparently the newt population had been increasing and they felt the need for more habitat. Even in the face of this obvious environmental problem, rather than cut off trade and damage the economy, trade continues because so many powerful livelihoods depend on the income. The scientists were busy with theories about things like crushed inner ears so they haven't done any work that will help the situation. Eventually, areas like the whole Mississippi River Valley up the central part of the U.S. collapse under water. By then it was obvious that dry land, an element essential to human life, would eventually be entirely transformed into the shallow, estuarine habitat suitable for newts. One would expect the humans would take quick action to cut off trade with the newts, but the forces of free trade and powerful economic interests remain paramount. At the end, the fellow who first commercialized the newts bemoans his decision that led to the global catastrophe. All he wants is for the children to forgive him.

A horrible story in a way, and one that is a parody of the world we live in. Today much of science is targeted toward things that can be developed commercially and most of science disregards the catastrophic problems we have. For example, there is an amazing amount of work going into human fertility technology, cloning and genetic engineering of food plants and animals, and the development of new drugs. We're doing better in the science realm than Karel Čapek's people because we are funding science in things like global warming and the role of industrial compounds in loss of human fertility even though we are pretty much ignoring the results we don't want to hear. For example, the best climate science predicts increasing droughts, floods, and unstable weather patterns that promise to sabotage our health and our ability to produce ample food to feed ourselves and our children. Yet we Americans elected a president who didn't believe in global warming so his administration habitually deleted scientific findings about global warming, especially those related to impacts on human health, from the official government reports. These reports form the basis of government policies, which in turn control many elements of our infrastructure that drive the structural consumption discussed in Chapter 2. Thus, we have not made any major transitions away from the technologies and practices that are causing climate change. We have little choice but to continue to buy the products made from fossil fuels whose use drives global warming—electricity from coal, plastics, paper diapers, corn, fuel for our cars, and many other things.

The People of the Corn

If you're paying attention, you might have asked yourself, "Corn?" Yes, our single-minded reliance on corn is one of our I, me, now "newts" that's leading us toward disaster. Let's follow corn's story for a bit to see how our technology, economy, environment, and well-being are all tied into one nasty knot that makes it difficult to extricate ourselves from this system that is damaging us and the planet. Our corn system breaks every one of nature's rules: it does not rely on current solar income, waste does not become food, diversity is not allowed, and the solutions to problems are anything but subtle and sophisticated. In many ways, the production and use of corn is a rape of earth's systems that does not bring lasting benefits to any living thing except the few who have grown wealthy from the industrial food system.

In a wonderful book called *The Omnivore's Dilemma*[59], Michael Pollan describes the natural history of three meals, one of which comes in two flavors: the first meal is from the industrial food complex and can be chosen with pesticides or without (e.g.,

---

59 Michael Pollan, "The Omnivore's Dilemma: A Natural Hisory of Four Meals," 2006, The Penguin Press

McDonald's/Pop Tarts/Stouffer's or mass-produced organic food); the second is from a grass-based pastoral farm where humans are working within nature's sustainable food web; and the third is wild caught, where humans are just another species in nature seeking to meet their nutritional needs (e.g., wild mushrooms, sea food, and venison). Most food in the U.S. today comes from the industrial food complex that uses chemical fertilizers, pesticides, mass production, and highly processed ingredients. The industrial food complex is based on a style of food production (inappropriately dubbed the "Green Revolution") so some family farms are a part of it as well as absentee investors. The base of the industrial food chain is mainly industrial corn, "#2 corn from Iowa" as Michael Pollan puts it.

The proof of this comes from carbon, the most prevalent element in living organisms. Like all elements, carbon comes in different weights called isotopes. Some isotopes are radioactive and decay over time while others are stable. Because of their weight differences, stable isotopes can be tracked through trophic systems, including the US industrial agricultural system's products and into our bodies. Our dependence on #2 corn from Iowa is so pronounced that the stable isotope carbon signature of our bodies reflects the stable isotope carbon signature of industrial corn. I'm sure you're thinking to yourself that you don't eat much corn, but you do unless you've found a way to eat only food from a pastoral farm or the wild. Industrial corn is made into high fructose corn syrup (a key ingredient in soft drinks and many other processed foods). It is also fractionated into most of the ingredients you find on boxed food as well as into the vitamins many of us take to make sure we get our daily minimum requirements.

And it is made into feed for chickens, cows, fish, pigs, tunas, and turkeys—this isn't the natural food of any of these species but they can live on it under the controlled conditions of CAFOs (concentrated animal feeding operations)—running the corn through their bodies doesn't change the carbon signature. Of course, raising meat in this manner does pollute the land and water near the CAFOs and creates a horror story of misery for the animals. This aspect of CAFOs has been captured in the award winning flash animation movie, *The Meatrix*, that can be found at http://www.themeatrix.com/. This little film is a parody of the popular *Matrix* movies and it exposes the lie behind our idyllic view of the family farm that has been transformed without our knowledge into a meat-making industry to the detriment of the soil, water, and the animals. For example, chicken beaks and toe nails are cut off to keep the animals from harming one another in their tiny cages and CAFOs are regularly given a free pass on Clean Water Act regulations when their manure lagoons regularly overflow into adjacent streams. The land is so degraded by the concentration of animal feces and disease that the operations have to be moved periodically. The popular excuse given for this travesty is "The only way we can feed our growing population inexpensive food is by using these industrial techniques". Unfortunately, the US Department of Agriculture, the EPA, and the US Congress have bought into the lie. The resultant policies, most of which have been sculpted by agribusiness interests, are a major factor in the steady loss of traditional family farms. But let's get back to corn because corn forms the base of the food pyramid that is our modern food system.

Corn is subsidized heavily by federal dollars through farm subsidies, mainly in the form of payments for fertilizer, pesticides, crop losses, and guaranteed prices when oversupply occurs. Corn is a crop that depletes the soil's nitrogen (essential for plant growth), a problem that used to be addressed by alternating corn crops with soybeans in the fields and by replenishing nutrients in the soil by adding cow manure to it. But that's not how it's done now. After WWII, the U.S. found itself with excess supplies of chemicals to make explosives and poison gases. Using the adage "Waste not, want not," government agencies promoted the use of ammonium nitrate as fertilizer and poison gas components as pesticides. This action did two things: it stopped crop rotation and also stopped the practice of spreading cow manure on the fields—suddenly soil fertility could be purchased with a government subsidy rather than maintained through traditional and reasonable care for the soil. Oddly, this chemical system is called "Green Agriculture." Now cows could be raised in feedlots rather than on individual farms—farmers who bought whole hog (pun intended) into Green Agriculture were free to go to Florida for the winter rather than stay home to take care of the cows and pigs.

With some genetic modifications, corn would now grow in massive, single-genetic-strain, crowded corn cities, doing the key transformational step nature requires in any food web of taking CO_2 from the air, water and nutrients from the soil, and using the sun's energy to make high energy carbon compounds. Producing food was taken out of the hands of farmers who had a vested interest in maintaining the health of their soil and put into the hands of an industrial system that looked at agriculture as a way to extract ever-increasing profits from each acre of land. The price we pay for this system goes far beyond the farm subsidies we pay.

Soil Health Is a Commons

I used to think soil was just dirt, but it's not. Soil scientists will tell you it takes 3,000-5,000 years to make a soil. Soil is a living system, full of worms, bacteria, microbes, layers of carbonate, and filled with minerals and various carbon compounds. Left undisturbed, nature builds soil through a succession of different plant communities that are able to grow at a given time—each generation dies and leaves its decaying bodies as nutrients that help build the soil. Whole communities of microorganisms live out their lives within soil. Without healthy soil, plants won't grow and an area becomes "desertified." That is, the living soil turns into dead dirt that won't hold water and lacks the trace nutrients plants need to grow.

The Great Plains of the U.S. lost a great deal of its soil in the Dust Bowl of the 1930s. These soils had developed over approximately 50 million years through the action of communities dominated by prairie grasses and buffalo herds. After the Dust Bowl, the U.S. formed Soil Conservation Districts (notice this is a definite social arrangement to control abuse) because we came to understand our dependence on this bounty from past ecosystems. Wouldn't you think the Dust Bowl and its aftermath would have taught us a permanent lesson? Instead, the transformation of the Great Plains farms into industrial entities founded on fundamental abuses of the soil has become the lynchpin of most U.S. food production today. Soil isn't the only thing that's a problem because plants won't grow without water.

Fossil Water Is a Commons

There is a vast reservoir of fossil water under the Great Plains called the Ogallala Aquifer that extends from South Dakota through the pan handle of Texas. With the discovery of plentiful supplies of natural gas (another source of stored energy from past photosynthesis), water could be pumped into giant center-pivot irrigation systems. These systems ensured that the corn in the corn cities would never be short of water, regardless of whether it rained or not. So the system for mass-producing corn was all finally in place. But as they say in Kansas, "The party just got started when the ice cream began to melt."

The water level in the Ogallala Aquifer goes down much more each year than it is replenished by rain. The depth of the aquifer varies but in the North Plains Ground Water Conservation District where it is shallowest, the aquifer is declining 1.74 ft/yr or about 10% per year. Farther south in parts of Kansas, the aquifer declines about 15 feet/year and is replenished with rainfall only a few inches per year. In the most severe cases, the Ogallala is losing up to 40 ft/yr. No one knows when it will be all gone. Remember Easter Island? In the face of the decline of the giant palm trees, the people substituted other things. Efforts are underway to conserve this irreplaceable water supply, but in the face of the economic engine driving the decline it will be impossible to stop the decline completely until water use is limited to the yearly supply from rainfall. I hope you recognize the tragedy of the commons in this situation. The time, species, global solution will have to be of the sort shown to be effective in ocean fisheries—stakeholders who sacrifice to replenish the aquifer would be the only ones to profit when the aquifer recovers.

In the meantime, this plentiful supply of water isn't really so plentiful and is being depleted along with the soil fertility. All the ice cream isn't melted yet because most wells aren't dry so the party keeps going. With the industrial agriculture system, nature has been automated as much as possible and the profits keep rolling in for agribusiness.

Most US citizens see agribusiness only in the form of supermarkets where we buy our food. But behind supermarkets, there is a system that involves many interconnected segments of the US economy, from corn production, to distribution, to processing and fractionating corn components, to various corporations that make soft drinks or diapers or prepared meals, to packaging firms, to marketers, to more distributers, and finally to the supermarket, where we are able to buy cheap food, a key thing we demand. According to Raj Patel[60], supermarkets are not just grocery stores. These self-service wonderlands of thousands of choices are designed to stimulate us to buy and consume. Odors of baking bread, roasting chicken, and smells of non-food items are wafted at us because consumer research shows we buy more when our sense of smell is stimulated. Behind the placement of items in the supermarket there is a fierce competition for particular spaces on the shelves. Check out products placed at a child's eye level versus those placed at your own eye level. A trip to buy some milk is actually a trip through a marketing maze designed to stimulate you to buy as much as possible before you leave the store. Everything we demand is there and much more.

The hidden worm in the apple is our ability to eat in the future when these commons are destroyed. But wait, there is one more glitch in the system—to ensure maximum profit, nature must be automated yet another step.

---

60 Patel, Raj. 2007. "Stuffed and Starved: The Hidden Battle For The World Food System". Melville House Printing, Brooklyn, NY

Industrialized Meat-making

Cows can't reproduce in feed lots—calves are produced by cattle that roam free on ranches, many in the mountains of Colorado on National Forest property with lifetime leases to particular ranchers. The problem of allowing the cattle to mate has been taken care of by the use of artificial insemination. One bull (who sadly for him never sees any cows) can father a whole generation of new calves. Still, this arrangement can be automated farther. Why not close the loop and automate the whole process of beef-making? Why not eliminate the bull? Recently the USDA approved the production of clones to replace the messy business of moving cows from one place to another. Now calves are kept in small plastic pens, identical clones of one another in identical enclosures with automated feeding. Profit can be measured in dollars per animal unit. This new mass production of beef won't be labeled as cloned meat because the USDA has determined that it is no different than beef already on the market. They weren't kidding about the "no different"—clones are all genetically identical to one another! This new breakthrough makes it so cows can be raised from beginning to end on corn even though their natural food is grass. So now fossil fuel in the form of subsidized corn can be transformed into steaks and profits.

To help his readers understand the meat production system, in *The Omnivore's Dilemma* Pollan describes how he bought one calf and followed its life through the process of meat-making. At one point Pollan visited his steer in a feedlot. The animal was living in a crowded space with thousands of other cattle, its eyes were red from the ammonia fumes coming from the slurry of fecal material and urine underfoot, and it was growing and gaining weight at 2-3 times faster than it would have grown on a grass diet. Pollan reported that these cows have acid stomachs that cause constant belching because although they can live on corn, it doesn't set well with them as a food. As a matter of fact, they can be kept alive only for a limited time on a corn diet because they get a disease called acidosis. Their complicated stomachs that are adapted to process grass become so acid on a corn diet that it eventually kills the animals. The trick is to slaughter them after they've gained as much weight as possible but before they die from the acidosis.

It struck me when Pollan described his miserable cow that it looked a lot like most U.S. citizens today as we eat our corn diet of chicken, turkey, beef, pork, milk, cheese, eggs, high fructose corn syrup, corn oil, and highly processed fractions of corn put together in various combinations to make Pop Tarts and other tasty treats. We have trouble with our weight and many of us suffer with acid stomach symptoms—notice the array of over-the-counter products that are anti-gas and antacid as well as the plethora of prescription drugs to curb "acid reflux." We've already looked at the economic sector associated with diet and weight loss products. Not a comfortable life we have, huh?

When is the last time you had chicken that tasted like chicken? Or milk that has some flavor? Or a steak that made your mouth water? The tasteless, abundant, cheap food that is the product of the industrial food system isn't edible in my opinion unless there is no other option. Others seem to share my opinion. In Denver, Colorado, there is a chain of supermarkets called Wild Oats that sell grass-fed beef and lamb as well as milk and other products from grass-fed cows. Those products are gone from the shelves before people will buy the agribusiness products. Good food should nourish our bodies as well as our spirits. It seems to me our industrial food system does neither. It doesn't follow nature's operational rules either: diversity is not allowed, current solar income isn't used, waste builds up and becomes pollution, and brute force is used to control problems.

Beef and Global Warming

So back to the original issue—how does all of this tie to global warming? According to Pollan, if you take everything into account from the feedstock to make fertilizer to the fossil fuel used to water, harvest, and process the corn, it costs ¼-⅓ of a gallon of oil to produce every bushel of industrial corn. Put another way, at average yields, it takes about 50 gallons of oil per acre to produce corn with the modern "Green Agriculture" system. In terms of energy, it takes about 2 calories of fossil fuel to produce 1 calorie of corn; this ratio gets lower yet when the corn is processed through a cow, pig, or chicken to "make" meat. Pollan says it takes ~35 gallons of oil to fatten one steer. When all the processing and transport is done, it takes about 10 calories of fossil fuel to produce one calorie of food today in the U.S. Before the Green Agriculture Revolution, it took one calorie of mostly renewable energy to produce two calories of food. Poor nations use one calorie of renewable energy to produce 10 calories of food. So essentially we have had a shift from farming—working with nature to produce food—to agribusiness—extracting dollars out of the land by inputting cheap, subsidized fossil fuel. And we are rapidly exporting this calorically backward system to poor nations so they can join with us in our insanity.

To put the magnitude of the damage of the Green Revolution into perspective we can translate the gallons of oil used into CO_2 emitted to the atmosphere. Of course the exact numbers vary depending on whether the oil has been made into fertilizer, gas, or whatever, and the formulas used are complicated. Nevertheless, it's instructive to look at some rough numbers. Each gallon of oil used puts ~22.6 lbs of CO_2 into the atmosphere. That doesn't sound like much, does it? But if we translate this into 6-ton Chevy pickup trucks, we float the carbon equivalent of one more pickup into the atmosphere for every 531 gallons of oil used. For every 15 steers we fatten in feedlots, we float another Chevy. No wonder the climate scientists say the "sky is falling."

Ocean Damage and Other Unintended Consequences

Remember William McDonough's reverse design assignment (chapter 4)? McDonough's point is that human design should be done with purpose and intentionality. If not, we suffer unintended and unforeseen consequences. Our agricultural system is bringing us a host of unintended consequences in addition to global warming. Agriculture producers figure if 10 lbs of fertilizer per acre will bring them a good crop, 20 lbs per acre should cause it to be a bumper crop, especially since the government is paying for the fertilizer. Excess fertilizer on the land seeps into water supplies so now the water in many parts of the Great Plains isn't drinkable because of the nitrates in it. For example, a March 24, 2009 utility warning letter told people from Hiawatha, Kansas, that permissible state and federal levels of nitrate have been exceeded and that the pollutant cannot be removed by boiling the water or letting it stand. They give a strong warning to avoid giving it to infants. Excess fertilizer also runs off the land during rain storms, flows to creeks, and finally enters the Mississippi River, which flows into the Gulf of Mexico. At the mouth of the Mississippi River and extending west towards New Orleans, the Gulf of Mexico has a huge dead zone where the ocean has been over fertilized to the point where no living things can exist. Think of the livelihoods of the fisher people who used to make a living along the coast catching shrimp and various types of fish and shell fish. The subsidized agricultural practices of the Midwest are killing the fisheries in the ocean. This isn't restricted to the Mississippi—around the world where American "Green Agriculture" has been adopted, near-shore ocean areas that form the nurseries where most edible ocean creatures reproduce are being killed by pollutants. This has a double whammy—a reduced food supply and the loss of the meadows of tiny plants in the ocean that pull CO_2 out of the atmosphere.

Monocropping, mass production of food, and transport of food and food animals from country-to-country has brought us devastating environmental and social problems, many of them occurring quietly while our Rome is burning. Uniform strains of food species that require particular fertilizers and/or feedstock supplant local genetic strains that have evolved in each place so diversity and stability is lost. I've read that a Caesar salad now travels an average of 2,500 miles to your table—no fire visible, but certainly another indication of Rome burning. Fossil fuels are used to transport food into regions that otherwise could produce their own food. Instead of producing food for their neighbors and themselves, local farmers have become cogs in the wheel of the industrial agriculture system. Jobs and family farms are lost. Use of fossil fuel becomes a necessary ingredient in getting food, which exacerbates the problem of global warming. Outbreaks of diseases like mad cow, bird flu, and foot-and-mouth are occurring with regularity. Diseases that previously occurred in only restricted areas now can move around the globe freely as livestock is bought, sold, and transported within countries and on global markets. Tainted food is unknowingly dispersed through a variety of products that are widely distributed and sold in supermarkets causing repeated epidemics of food poisoning. How secure is our food supply under these conditions? Not very because we aren't following the operational rules of nature as we've turned food production into a system that treats food production like an assembly line in a car factory.

In return for our acceptance of these unintended consequences, our industrialized food system has released the majority of our population from the work of raising and preserving food to manufacture and sell consumer goods. We also get the cheap food we demand. For example, in 1989 Americans spent less than 10% of their income on food whereas people from India spent more than 50%[61]. This sounds like a boon for Americans with an I, me, now perspective because the lower food costs free up our income to buy more consumer goods that are assumed to improve the quality of our lives. With freedom from living near our food supply underwritten by the use of vast stores of fossil energy, we have developed "The American Dream," which more correctly should have been called "The American Nightmare" because this food production system has done more to separate us from our environment and our roots as biological organisms than any other aspect of industrialization. The balance between our activities and water availability in each local place is a prime example of the consequences of this disconnect.

---

61 Traub, Larry. 1992. Food Review. "Per Capita Food Expenditures Declining Around the World". http://findarticles.com/p/articles/mi_m3765/is_n1_v15/ai_13608617/?tag=content;col1

Water Limits

In the early part of the 20th century, supplies of water were a limiting factor determining which activities occurred where. For example, manufacturing was concentrated around the Great Lakes because at that time dilution-was-the-solution-to-pollution and pollution was a necessary byproduct of industrial activity. Our modern legacy from that practice is that the Great Lakes region of the U.S. is the center of industrial manufacturing in the US and the region has more superfund sites than anywhere else. The abundant water in the Great Lakes region is polluted water—strict advisories are in effect for eating fish caught in the region because the bodies of fish concentrate deadly pollutants like PCBs. The effects from eating Great Lakes fish include loss of fertility, cancer, cognitive difficulties, diabetes, and a host of other problems. In addition to the chlorinated hydrocarbon-type of pollutant, the Great Lakes also harbor tons of high level nuclear waste, much of which is stored near these huge bodies of fresh water and all of which is 100% lethal to all forms of life. So things are rough in the Great Lakes, but at least Great Lakes people have plenty of water, even if it's polluted. There are other areas of the U.S. where fresh water is at a premium.

I lived in the San Luis Valley of south-central Colorado for 18 years. It was cold there at high elevation but the sun shone every day. We had plenty of water from the underground aquifer of fossil water that was replenished by snow melt from the surrounding mountains. I'll never forget one day shortly after I moved there. We were moving from Syracuse, New York, where the asphalt sidewalks floated during heavy rainstorms. The San Luis Valley gets only about 7 inches of precipitation per year so this was a great move for a person who had a couple of green iguanas whose happiness and health depended on sunshine for basking. The poor things can't even digest their food unless they have an outside source of heat and, if they are to remain healthy, ideally the heat must come from sunshine. Anyway, it's wonderful to live in a spot where the sun shines every day and you don't have to constantly dig out from under piles of snow. Still, snow is readily available for winter sports—Colorado is noted for its ski areas. Mountain snow feeds the supply of water to lower areas but even so water is at a premium in the San Luis Valley. I remember shortly after I moved there, two landowners in the nearby mountains got in a gun battle over water—one of them was killed. I thought I had really moved into the "Wild West."

In today's "Wild West" water is money. The farmers of the San Luis Valley used to get up in arms claiming that the Texans were seeding the clouds and stealing the rain. At least the farms in the San Luis Valley can still grow a crop during drought times because the land is set up for irrigation using the fossil water in the aquifer below the valley floor. The water in this aquifer is so plentiful artesian water comes to the surface in many places, much of it heated by geothermal energy. This water is worth lots of money—rights to it are controlled tightly and the most valuable water rights are from the old Spanish land grants because during a shortage they have first rights. Notice the definite social arrangement to protect a commons? Not all of the Wild West is so organized. In the west water doesn't flow downhill, it flows toward money.

By the late 1980s, a Canadian company had noticed the San Luis Valley's water and quietly bought up a right-of-way from the valley to Denver, a five-hour drive over a six mountain passes. They also bought the water rights to the Baca Ranch near the Great Sand Dunes National Monument for several million dollars, land that was an old Spanish land grant with its first rights to available water, which, of course, was communally held in the aquifer. Their plan was to pump water out of the San Luis Valley and sell it to the Denver area—the value in the 1980s was in the vicinity of $5 billion per year. Overuse of the aquifer and harm to the valley, its farmers, and the non-human species wasn't their concern. I'm proud to say that the people of the San Luis Valley banded together and voted 97% in favor of taxing themselves to fight this theft in court. Fortunately (or unfortunately depending on your perspective), there was a previous example from the Owens Valley of California. That lush valley became a wasteland when it was drained of its fossil water to supply the people of Los Angeles. Armed with the Owens Valley example and careful studies of the geology of the San Luis Valley, the valley people won the fight against the Canadian company, but the issue will come up again because the water situation in the west is getting worse each year. I understand the Baca is now under the control of the Great Sand Dunes National Monument so the next effort to steal the valley's water will have to have another strategy. The story of the San Luis Valley's water is one bright spot in an otherwise grim picture of water flowing toward money.

Much of the problem lies in the disconnection between people and the land. The Western population of the U.S. rose from 4.1 to 69.4 million people between 1900 and 2006[62]. This was a period of mostly wetter-than-average times so water diversions could be used to support the myth that water is abundant in the dry West. Until recently the fastest growing population

---

62 Robert Kunzig, "Drying of the West," National Geographic, February 2008

in the U.S. was in Las Vegas where the per capita water use is the highest in the U.S. People have huge private pools, irrigated golf courses abound, and misters spray constantly at outdoor eateries to keep people cool. Agriculture uses a great deal of water too—the Imperial Valley of California is irrigated to produce the lettuce, strawberries, and other fresh produce we enjoy all winter. There's a cost to this though because the once mighty Colorado River that in the past dug the Grand Canyon now no longer reaches the Sea of Cortez adjacent to Baja, California. Instead, the Colorado River is now partitioned between states feeding the ever increasing demands of our civilization. The sad part of the story is that our own actions are undermining the systems that supported this growth.

Each water diversion robs whole valleys and the species living there of their water. The full impacts aren't immediately apparent but as the western climate dries, more and more species come under stress or disappear. Early climate models predicted global warming would dry the west. This is coming true in spades. Snow pack is declining and soils in the mountains have become so dry in places they won't absorb water. In some areas, the trees have been weakened and have been invaded by beetles, making them into dry tinder that fuels uncontrollable forest fires. After the trees are burned, erosion of the fragile soils follows, and rain then causes devastating mud slides.

Collapsing Civilizations

Mystical thinking that we can somehow engineer ourselves out of this imbalance is wrong. The Anasazi of the four corners region (Colorado, New Mexico, Utah, Arizona) had a flourishing culture about 1,000 years ago—they disappeared when a prolonged drought hit the southwest about 1130 A.D. It took only 30 years to crash their civilization. Why would we think our civilization could do any better at persisting when we are systematically destroying the foundations of water and soil upon which we depend? This is especially dangerous when you consider we are rapidly exporting our civilization's habits as we globalize trade. Globalization means not only new trading partners but changes in food, water, and soil use around the world as unwitting countries adopt the superficially successful American model of prosperity.

China has been the poster child for "astonishing economic growth" in the past two decades[63]; its GDP has increased by 8+% on average each year. China is also one of the oldest civilizations on earth—their basic relationship with each other and the earth hasn't changed appreciably during that time. They have not joined the ranks of the hundreds of civilizations now extinct so they must have been doing something right, at least until recently. Under the influence of our civilization's trademark free market economy and reduced trade barriers, China is now adopting more and more of the U.S.'s way of life. The Chinese, the world's largest population, are rapidly becoming urbanized and separated from their food production. Concurrent with this, they are eating fewer staples like corn, rice, and wheat while more water-intensive foods like fruits and vegetables have increased in their diets by a factor of more than four since 1960. They are also rapidly increasing their consumption of meat, a change that can be measured by food-related per capita water consumption. Cereal crops use 0.84-1.3 cubic meters of water per kilogram of yield, but beef takes ~12.6 m^3 per kilogram

---

63 William J. Baumol, Robert E. Litan, Carl J. Schramm, "Good Capitalism, Bad Capitalism" 2007, Yale University Press

of beef. Even with these changes, China's food-related water consumption per capita is still only about ¼ as large as the U.S.—we use about 3,074 m^3 per person per year[64]. Much of China is like the U.S. West so these trends project into the need for importing large amounts of food to China in the future. This will help the balance of trade, but it sure won't help the balance of nature or the future of the Chinese civilization. The Chinese have joined us in our rush to disaster.

Local Food, Local Water, Local Collaboration

Food in nature is a local phenomenon, just as evolution is a local phenomenon. Water is also a local phenomenon. You probably haven't thought about it, but even our bodies use water temporarily and then put it back into the area we live. Drink-pee-drink-pee, etc. In this local, natural world, nutrients cycle in the ecosystem, and food energy is transmitted through food webs of communities of organisms living together and underpinned by food energy produced through the capture of the sun's energy in the process of photosynthesis. There is a great deal of diversity—species are interspersed among one another so disease-causing organisms can't spread easily between individuals. Collaborations are established between species to ensure adequate water and nutrition for both. For example, funguses frequently coat the roots of other species in a symbiosis that helps plants absorb water and nutrients more effectively while the fungus derives its food from the nematodes that otherwise would have attacked the roots. Species in the communities even collaborate with one another to help ward off predators.

---

64 "A thirst for Meat," Science News, January 19, 2008, reporting on work by Junguo Liu and Herbert H.G. Savenije, Hydrology and Earth System Sciences.

One day I was working at my computer, which is located by two big windows, one facing north and the other east. It was summer so the windows were open, the sun was shining, and I could hear the birds going about the business of feeding their newly hatched or fledgling young. All of a sudden, I was startled by what sounded like an explosion in the birds outside. I could see a multi-species flock of birds—chickadees, nuthatches, blue birds, robins, and others—flocking together and chasing a crow. My neighborhood sticks to songbirds—there aren't many nasty species like crows with their raucous calls. But a crow had swooped in and grabbed a baby robin. I ran outside and by then everything had quieted down. All that was left was a few baby robin feathers. I was amazed. I hadn't realized that the birds living around here had a "neighborhood watch association". It's easy to see that this behavior could easily evolve because all of the individuals and species involved would have had better success at protecting their babies when they banded together.

The only problem with the system would be the cheater birds that take advantage of the protection of the commons but who don't help their neighbors. In humans, local cooperation in food production and consumption as well as local cooperation in soil and water conservation is just as important as local cooperation in educating our children and protecting ourselves from fire and outlaws. The American industrialized food system, which takes without replenishing and damages without repairing, is a cheater rather than a cooperator in the natural systems of earth and a "crow" in our local communities.

Our unwitting acceptance of this ubiquitous cheater—the industrial agricultural system—in our midst has left us facing a multitude of environmental problems as well as severe health problems. By any ethical measure, behavior promoting this sort of exploitation is not acceptable. It's also immoral to damage our children and their future just for our own convenience and economy. We must recognize the drivers of this devastation are our insistence on cheap food along with our rush to feed our I, me, now "wants" within a system that rewards profit now regardless of the time, species, global costs. The rules of nature—waste equals food, run on current solar income, thrive on diversity, use information-rich solutions—are all defied by our current agricultural and water use systems. We've sold our lives to cheap food and artificially abundant water without regard to the deferred costs. But the debts are coming due, including the risk of global famine and filthy water. Looked at another way, the true cost of our current food and water practices are too high for anyone to afford. The sad thing is that we don't have to do things this way. We're living with the results of some frozen accidents in our evolving relationship with soil and water. Fortunately, these accidents can be undone if we begin to think ethically for this time and place.

Changing our behavior doesn't have to be devastating and it doesn't have to mean we must live in a state of deprivation to "save nature." Life on earth is naturally abundant, even in spots we wish were less abundant. Think about your dish rag—in short order all sorts of living things are growing on it making it have that stale smell. Because of the evolutionary processes—excess reproduction, variation, and survival of the fittest—the living world is one of abundance. What we have to do is to modify our activities so we leave a wake of health and abundance behind our activities rather than depleted soil, polluted water, and loss of genetic diversity. If we're clever about how we go about food production, we can still have relatively inexpensive food, free time to pursue other economic goals, improve our food security, and have a quality of food that does nourish our bodies correctly as well as our spirits.

Natural Abundance and Biodiversity

Bill Moyers' television special *Earth on Edge*, an early report on the United Nations' Millennium Ecosystem Assessment Project, looked at biologists who are going through the global household taking stock of how things are. Two pieces of this special that concern food and water are important to this chapter—one is on Charlie Melander, a farmer in Kansas who has developed a system that farms his soil rather than farming government subsidies; the other is about water in parched South Africa.

Gary "Charlie" Melander and his wife Kathy sharecrop ~2,500 acres in central Kansas growing wheat, grain sorghum (milo), and soybeans. Charlie has developed a combination light till/no till system that reduces fuel, herbicide, and pesticide use saving $2.00/acre on fuel and $21/acre on herbicides[65]. To control weeds, conventional till systems plow the soil deep and no-till systems require gallons on herbicides. Deep tilling loosens the soil, makes conditions excellent for weed growth along with the crop, and exposes the soil to wind erosion. Rather than doing either exclusively, Charlie does what the field needs and what works to produce the best crop. He's found he can plant wheat in milo-stubble without tilling and without herbicides. The milo residue keeps weeds from growing, the soil is protected from erosion, and the crop rotation keeps insect pests from becoming a problem. Charlie and Kathy also keep cattle

---

65 Gary "Charlie" Melander. "A Farmer Looks at Energy Efficiency," Presented at the ACEEE Forum on Energy Efficiency in Agriculture, November 16, 2005, Des Moines, Iowa.

that are allowed to eat the residue in the fields, increasing soil fertility naturally and giving them another farm product to sell. They've put in wind rows of the native deep-rooted prairie grasses to provide habitat for prairie species, many of whom help keep their crops from being damaged by pests.

When the no-till fields would normally be coated with herbicides, Charlie does what he calls minimal tilling. Using modern farm machinery, which has huge disks with hydraulic controls, he tills to a shallow depth, thus protecting the soil stratification and turning up big soil clods that suppress weeds. No soil blows off of his fields after he's tilled them—as a matter of fact, his method collects the wind-blown soil from his neighbors who pulverize their soil to tiny, light particles that blow easily. In short, rather than spread herbicides, Charlie is using big dirt clods. In the midst of his neighbors who have become agribusiness men, Charlie is using nature's organizational principles to his and earth's advantage. He is promoting genetic diversity, reducing his reliance on energy outside of the sun's energy, waste becomes a profit center, and he is using information rather than brute force to solve problems.

There are many benefits to Charlie's clever way of farming. First, his crop yield is as high as or higher than his neighbors who farm conventionally—we won't starve or have food shortages if current agricultural practices are changed. In addition, his production costs are much lower than his neighbors, a fact lost in our current system because of the way our farm subsidies work. His neighbors' expenses are paid for by the government—that's why the conventional system should be called "farming the government." Charlie saves $2.00/acre in fuel costs and $21/acre in herbicide costs. He isn't contributing to the water pollution that makes much of Kansas' water undrinkable and he isn't contributing to the dead zone in the Gulf of Mexico. His farm is an oasis for many different species, most of whom help him control insect and weed pests. Lastly, Charlie's system doesn't require massive amounts of irrigation to grow his crops. Deep-rooted plants near his fields capture rainwater and channel it deep into the soil. The no-till and clods from minimum till leave a soil surface that retards water loss from the soil, much as mulching does for water absorption and retention in ornamental plantings. So another big benefit is water conservation and maximal use of available rain. It makes so much sense that I have to wonder why all the farmers in Kansas aren't following Charlie's lead.

One reason is that the agriculture branches of the universities aren't interested in hearing what Charlie and Kathy have discovered. The agriculture scientists are like Karel Čapek's scientists—rather than being open to lines of investigation that take the bigger picture into account, both sets of scientists work to support or enhance the current dominant system. In his paper presented at the ACEEE Forum on Energy Efficiency in Agriculture, Charlie Melander says: "Farmer Jones plays crop lottery: double up on fertilizer tickets, pray for extra rain, and wait to hit the crop jackpot covered with subsidy dollars. What do the taxpayers get for their subsidy dollar? Artificially cheap farm products in the supermarket, fewer farms, falling water tables, nitrates in the drinking glass, and a dead zone in the Gulf of Mexico. These frenzied maximum input, maximum production farm models, endorsed by the universities, the large agriculture corporations, and government are insanity. The taxpayer and the country deserve much more from their subsidy dollars."

Charlie's solutions are to create government policies that cause "Farmer Jones to collect more subsidy dollars by using environmental tactics that reduce input costs." Such a change would radically alter the dynamics of our agricultural system, the quality of the food we can buy, and the sustainability of our food supply. Of course, the oil and chemical companies would fight changes of this sort, but after all who is in charge here? A few CEOs with tons of money they collect from us? Or are we in control? After all is said and done, we are the ones with the votes and we are the ones who must supply them with money. Isn't it time we recognize the power we have to change things?

Natural Water Abundance

Bill Moyer's television special also had great information about water. I love this story because of the "tree-hugger" stereotype of environmentalists. In South Africa there are 40,000 people working for water by destroying non-native trees. Apparently the early Europeans didn't like the look of the land because the rich array of species—20,000 strong—are mainly water-sipping, low-lying plants. One area, the fynbos vegetation zone in the Cape, is so unique that it is one of the six floral Kingdoms of the earth. The Europeans spread seeds of European trees, many of which siphon water out of the soil and dry out the streams. The stream flow in southern South Africa has been dropping by halves about every 25 years, partly because the trees have been reproducing happily in their new habitat. So the South Africans have hired their poor, mainly Kohsian people, to work on teams to remove the non-native trees from the land. This is one problem that has produced two solutions—gainful employment for the poor and an increased supply of water that allows economic development to proceed. The lessons to be learned here are that local water abundance depends on protecting the natural water systems that have developed during the geological history of each place and the best way to solve water shortages is to live within the water budget of each watershed or underground aquifer.

Today the water problems of various regions are the product of bad thinking and bad technology and a lack of adaptation to differing ecologies. Agriculture uses 40% of freshwater withdrawals in the U.S and energy generation uses another 40% (coal and nuclear plants create steam to drive steam turbines that generate electricity). The remaining 20% is used directly in our homes and businesses[66]. The agricultural, energy, and domestic technologies we use in the arid southwest are the same as those used in the water-rich Great Lakes region. According Mike Hightower and Suzanne Pierce in an article in the premier journal *Nature*[67], in the face of a looming crisis over water shortages the U.S. lacks a coherent, integrated plan for this most vital of all resources. In the United States, more than 20 different federal agencies are responsible for various aspects of water policy. Frequently the policies are at odds with one another. For example, environmental regulations severely restrict the re-use of domestic waste water that is now normally mixed with dirty toilet water so we are not allowed to use our waste water to irrigate our lawns and gardens. At the same time, the abundant solar energy supply of the arid southwest is ignored

---

66 Nature Editorial. 2008. "A fresh approach to water". Nature 452:253. March 20, 2008.

67 Hightower, Mike, and Suzanne A. Pierce. 2008. "The energy challenge". Nature 452:285. March 20, 2008.

as an energy source while water-hungry coal-fired and nuclear generating plants are used to produce electricity.

Meanwhile government policies allow the diversion of the water of entire drainage basins to feed the growing water demands of those living beyond their place's water budget while ignoring the natural abundance within the water budget of their place. One day I went to work with my brother, Dayle, who is a civil engineer working mainly on water projects in Colorado. As a normal part of his work, Dayle must go to the Denver Water Board frequently to get permits for some of the work he does. On that day, I was astonished to see a huge real-time, lit display of water diversions occurring all over the mountains of Colorado to feed the water needs of Denver. The Continental Divide runs down the Rocky Mountains; water that falls on the Western Slope runs toward the Pacific Ocean supporting all of the local places in between and precipitation that falls on the Eastern Slope flows toward the Mississippi River and the Gulf of Mexico. Nevermind that. It looked like all of it flowed toward the cities along the Front Range of Colorado: Boulder, Denver, Colorado Springs, and so on. Precipitation and ground water in the Front Range of Colorado supports short grass prairie, countless birds, and Pronghorn Antelope. Local ponds support an abundance of aquatic species and the large rivers flowing through the cities have naturally wet lowlands. If the people living in this region were xeriscaping (dry landscaping) their yards and using water-saving devices like low-flush or composting toilets in all bathrooms and if it was mandatory that waste water would be used and re-used, then it wouldn't be necessary to drain the northern Rocky Mountains to meet the needs of the people in Denver.

Those who live in a desert need to be actually living in their place, not in a world of lies built on artificial abundances of scarce water. Those who must have a lush green lawn should live in places where those plants will grow without irrigation. If you stop to think about it, it is an insane idea to irrigate deserts to grow turf grass and golf courses when our major rivers can't reach the ocean any more. If we approach problems with a time, species, global perspective and make ethical choices appropriate for this time and place, we do not need to face the oil, water, and soil shortages that loom now and in the future of our present course.

We have to be very careful of what we pay for with tax dollars so behaviors that fit within the rules of nature are rewarded and those that defy those rules are discouraged. That means that we should not be subsidizing "Green Agriculture" farming or research, nor should we be spending tax dollars on mass-production or marketing of agricultural products. The commons of our soil and water must be protected because without these commons, we cannot live. We should subsidize farming practices that build soil, increase diversity, and conserve water. We should not be separating more storm sewers from sewerage and domestic waste water should not be mixed with sewerage. We should demand on-site water management in our zoning laws. We should not be funding any more water diversion projects out West. We should fund research into cultivation methods and companion plantings that are drought resistant. And the list goes on, limited only by our amazing store of human creativity.

Important Concepts of this Chapter

Soil and water are vital commons that underpin our well being yet these commons are being decimated by our modern food production system that has inappropriately been labeled "Green Agriculture".

US Green Agriculture techniques are being exported to countries around the world where they create the same social, economic, and environmental problems they create in the US.

Fossil-fuel-intensive food production systems based on monocrops of corn and corn products contribute heavily to global warming, loss of biodiversity, depletion of soil resources, and depletion of water. This kind of food may be cheap to buy, but it is not healthy for us to eat and it is devastating to earth's stability and abundance.

Water is the life blood of all living organisms but we are treating it as a commercial commodity, fouling it with our current systems that turn precipitation into storm water, and failing to adapt our water-use behaviors to each place.

No matter what the cost, things aren't cheap unless the future price is affordable.

To be ethical, we must re-design our relationship with the earth as we go about the important activity of food production. This re-design must be centered around the operational principles of nature and must be designed so its selective pressures encourage protection of soil and water.

New thinking in agriculture and commercial development demonstrates we can increase our agricultural output and increase the abundance of clean water without damaging our lives, health, or prosperity. As a matter of fact, working with nature on our food and water abundance will assure us of abundance for all the generations to come.

* * *

Modern US land use practices are a heritage we have developed from our history and this is a broad issue impacting our lives and setting social norms for us that are no longer functional. A time, species, global view of land use traditions in the U.S. reveals a series of goals we have adopted that are no longer functional for us because they shape our behavior in opposition to nature's operational principles.

Chapter 8

Land Use Traditions

Between the founding of the U.S. and now, Americans have developed cultural norms related to land use. Remember the power of paradigms, deeply held assumptions, from "The Origin of American Bad Habits" in Chapter 4? We have some cultural habits related to land use, some of which are destructive to our lives and to nature because these habits are our definite social arrangements that were adopted to protect our common interests, many times with an I, me, now perspective. Unfortunately, our outdated land use habits no longer serve us well because they fly in the face of nature's organizational rules.

Broadly speaking, the land in the U.S. is all designated either as public or private land. The public land is owned by various levels of governmental agencies from small communities up to the federal government. On the other hand, private land is deeded to individual people or to various legal organizations, including non-profit and for-profit corporations. You may not have thought about it but in the U.S. every square inch of land has a property deed and is owned. That's not true of every country around the world and it wasn't true when Europeans first colonized America. All over the world, people live on the land but in many places they don't own it. The US system reminds me of a Gary Larson Far Side© cartoon from years ago. A couple of fleas are on a dog—the dog's hair looks like giant trees around the fleas—and the fleas obviously have set up ownership of the entire dog. Other cultures aren't necessarily like that. For example, in Brazil much of the land isn't deeded even though people may be living on it and using it. Think about the tribal lands of the Amazon natives that are now being deeded to mining companies, or to cattle and soybean operations, without tribal approval.

The concept of land ownership coupled with our heritage of definite social arrangements (laws) for it has created the world we drive through as we go to work, the supermarket, or on vacation. It has also created the system of public lands we hold dear in the US. Let's take a look at the historical origins of our current system.

Public Lands—Preservation and Conservation

In the U.S. we are especially concerned about wilderness but wilderness is a concept unfamiliar to many civilizations. It comes from the heritage of our ancestors who landed on this continent and then pushed back the "wilderness" to bring in "civilization." Never mind the Native Americans who lived as a part of the wilderness. Actually our ancestors considered them to be "wild" too, but the Native Americans had no concept of "wild." By the 1890 census the frontier in America was considered to be completely conquered. Americans were uneasy with the loss of the wilderness, so by the early 1900s, two schools of thought developed as foundational to much of our public policy today with respect to public lands. Each, preservation and conservation, has a strong place in our culture and together they represent the first environmental movement in our history.

Preservation thinking started with John Muir, a man of the mountains raised in poverty in a strictly Christian family, a founder of the Sierra Club, and a person who likened pristine wilderness to a temple. John Muir considered nature to be perfect as God made it; he thought the world wasn't made just for the use of people—it had value in and of its own right. Clearly, John Muir did not consider people and their activities to be a part of nature; in his view the natural world had to be protected from the activities of people, essentially preserved as it was found. This makes sense because John Muir lived during the early years of the industrial revolution when non-renewable resources were being scooped and scraped from the earth and made into consumables without concern for environmental impacts. John Muir's legacy to us is the National Parks, many of which were preserved by Republican presidents; notable among them was Teddy Roosevelt. The national parks have very strict rules for public use—if you go there, you leave only your footprints and they have to be light footprints—the resources in National Parks cannot be developed. National Parks are essentially a sanitized wilderness, areas where the Native Americans and most wolves and bears are removed. Preserved land is land with no permanent human occupants but also safe for people to enjoy the aesthetic wonders of "wilderness."

The concept of wilderness is so deeply buried in our psyches that we spread it around the world as we give aid to other countries. Americans contribute to organizations that buy pieces of rainforests around the world to set up nature preserves and our government supports the development of national parks in biodiversity hotspots in poor countries. Part of the motivation given is to develop these areas for ecotourism. In all cases, native people who had been living in those places for generations without degrading the ecosystem are relocated to areas outside of the reserve and encouraged to become nature guides or similar work that supports ecotourism. Wilderness in the American mind does not include people as residents, only as visitors. The assumption that people are not a part of nature (Chapter 4) is deeply embedded in our culture.

Very few, if any, poor people have historically had the privilege of enjoying nature preserved in the national parks because it has always been expensive to get to these remote places with the requisite equipment. The early unwritten goal was to preserve a mystical playground for those few wealthy enough to get to it. Unfortunately for the national parks, now there are an awful lot of people wealthy enough to travel; end-to-end RVs clog the roads in U.S. national parks almost every day all year every year. The national parks have become a commons that is being "loved" to death. Too few people have the money or the interest to visit the nature preserves established in poor countries so the relocated people have a hard time making a living with the loss of the ancestral lands.

Conservation, on the other hand, is the legacy we have received from the work of Gifford Pinchot, a child of wealthy parents and the first head of the U.S. Forest Service. As an aristocrat, Pinchot had never been denied access to things he wanted, an attitude apparent in conservation thinking that forms the basis of the laws associated with the bulk of public lands in the U.S. including the Bureau of Land Management (BLM) holdings, national forests, wilderness areas, and other public lands. These are multiple use areas—recreation, public lands ranching, hunting, fishing, logging, mining, and other uses are permitted. Pinchot had a utilitarian approach; the goal on these lands was to maximize the use of the land and its resources for as long as possible while accepting the decline of the ecosystems and the depletion of the natural resources.

Public lands don't bring in government revenue; as a matter of fact they cost taxpayer money to support them as a common good. In some cases, users pay nominal fees for access. For example, mining or logging or cattle grazing in national forests costs users a small portion of the profit they gain from their activities. This system is obviously in place because theoretically Americans can buy less expensive products if it costs the producer less to acquire the raw materials. The price of wood to build homes is less expensive because the supply is increased by logging national forests. The cost of beef is less because some ranchers have hereditary rights to graze their cattle in national forests thus saving most of the expense of feeding them as they reproduce and the calves grow. This logic doesn't include the price we pay as taxpayers to support public land or the price we pay as the commons of public land is degraded.

Still, public lands have served us well over the years. But if they are to continue serving us into the future, we must modify their use using a time, species, global perspective and what we know about protecting a commons. The overuse of national parks could be reduced by instituting a lottery system that allows only a certain number of people to go through. The price of resources from conservation lands could be increased to fair market value and withdrawals of resources could be limited to renewable levels. Destructive, polluting, invasive uses like mining should be prohibited because these uses are not sustainable. One way or the other, the public land of the U.S. is a jewel in our crown that should not be chipped away by misuse.

Private Land

Neither preservation nor conservation philosophies were used on private land—there the landowner was king. Early on in our history we protected the rights of private land ownership by imposing very few restrictions on what landowners could do with their private land. By the middle of the 20th century, we had gotten ourselves to a spot where our farming and industrial activities were so prevalent and damaging that whole species and ecosystems were disappearing. By the 1960s, chemical pollution of air, water, and soil had become so pronounced that biologist Rachel Carson's landmark book *Silent Spring* moved people to organize the first Earth Day. That movement, the second great expansion of our environmental awareness, gave rise to the Clean Air Act, the Clean Water Act, and the U.S. EPA to regulate the impacts of land owner activities on the common good. This expansion of the environmental movement was substantively different than that of the early 1900s.

With the origin of the EPA and corresponding state agencies, land owners were suddenly curtailed in their activities. Polluting activities could still occur, but now they could only be done within the limits of the law. Another way to state this is that it became legal to poison wildlife and your neighbor as long as either no one was aware of your activities or you limited your output of toxins to some legal limit. The legal limits are monitored by the EPA, and as long as the air quality or water quality doesn't drop below certain levels, just about anything goes. Swamps came under protection too because it was discovered that if individuals drained enough swamps and farmed or developed right up to stream edges, soon periodic floods caused great damage to everyone in the watershed and downstream water was so polluted as to be unusable. So landowners were still allowed to drain swamps—now they had to replace the wetland with other "comparable" wetland—never mind whether the species adapted to the old wetland were able to move into the new so man-made lakes and polluted golf course ponds soon took the place of nature's wetlands.

Through all this history and up to modern times, most private land use was regulated mainly by small communities within the framework of a few relatively nonintrusive state and federal laws. In Michigan where I live, the planning commissions of township governments control changes in the primary use of private land. Of course, state regulations are involved when water quality is directly impacted, like draining wetlands and diverting rivers. Still, the major control on whether land is used for agricultural purposes or given over for "development" is a township issue. Forest that is not in the public domain is subject to the whims of the landowners; it is their choice to keep their woodlot or to sell it for lumber.

In most of the U.S. each state is divided into counties and counties are divided into townships and municipalities for taxing purposes and various responsibilities for public services. Even 50 years ago, this system wasn't too bad because local control over local land use seemed to work for the small, rural, isolated communities that dotted the states. Now this system has evolved into an insane monster because the thousands of small, locally-correct decisions by this multitude of entities are adding up to a horrible disaster in the U.S.

This history of the evolution of our culture is the root of the policies, laws, and assumptions we are still using today. But today things are different—our population and the personal wealth of most Americans have increased dramatically, the legacy of pollution from the past has come back to haunt us in the form of toxins in our food, our bodies, and in our food animals, and forests and farmland are becoming more and more scarce. The system we use now grew out of an I, me, now perspective and ethical framework but times have changed. As we move forward into the future, we must move the foundational perspective of our personal behavior and public policies to a time, species, global perspective and function within nature's operational rules if we are to solve the great energy, pollution, social, and food problems of our time. Various perspectives are valuable here to understand the severity of the problems caused by our land use customs and laws.

Townships or Cities

Let's look at land use first from the perspective of a township or municipality. If they can attract development, then they increase their tax base because housing or strip mall developments pay more taxes than farms. Increasing development also brings more jobs to a community and increases the economic activity for all the businesses in the area. Notice the increased economic activity is measured in terms like GDP, not GPI discussed in Chapter 3. So loss of woodlots and loss of local ecosystem services like insects that control pests or pollinators that increase crop yields are not counted as a negative. The increased costs to taxpayers for schools, fire protection, and roads aren't counted and things like more wells pulling water from the aquifer of the area are ignored. Possibly most damaging is that the development permitted in one township isn't coordinated with what's occurring in adjacent townships, a factor that leads to a hodge-podge of development that makes no sense when you look down on the results from a bird's-eye view of the land.

Corporations

Let's look at the situation from another perspective. Corporations now are huge entities with the resources to move from state-to-state and nation-to-nation at a whim. Nowadays corporations frequently demand massive tax concessions in order to site their business in a particular township, county, or state. As they do this, they incite bidding wars between small, local, amateur entities in townships widely separated from one another. The net result is little townships giving millions of dollars of taxpayer subsidies to polluting industries that provide a few minimum wage jobs. For years, I've watched this occurring in the area where I live. In some instances we've paid hundreds of thousands of dollars in tax rebates for one job that pays minimum wage. It seems to me that small communities would be better off just hiring the person themselves—we could pay higher wages and save money and our environment at the same time.

One notable instance of what I'm describing occurred with the turkey slaughter industry that was sited in Borculo, a small town south of where I live. This company was slaughtering thousands of turkeys each day and the waste was overflowing into the Pigeon River, a river that flows through Pigeon Creek Park, the last place Passenger Pigeons were seen before the species became extinct. Anyway, finally the state Department of Environmental Quality set its foot down and insisted the company provide adequately for the waste. About the same time as the pollution events were curtailed, the company put out meat that was tainted with *Listeria*, a bacterial contamination that killed 19 people nationwide, mainly women and children. In the wake of these disasters, the company was sold to Sara Lee, a corporation that provides turkey and other meat to the airlines and other entities nationwide. In short order Sara Lee had pitted the tiny town of Borculo against a small town in Iowa in a bidding war of tax incentives—the prize at stake was whether Sara Lee would stay in Borculo or move to the

Iowa community. Is that insane or what? Using tax dollars to attract land uses that are destructive of the land. And all for a few minimum wage jobs.

The story gets worse. Those Sara Lee jobs are not desirable. Since I've lived here, one employee fell into a meat grinder. Needless to say, he didn't live through it. I'm not sure whether the meat made it to market or not. One way or the other, I know what I got out of the situation. Semi traffic past my home at all hours of the day and night, my taxes subsidizing the profits of this corporation, turkey waste in the local rivers, and turkey feathers all over the road.

States

Let's look at the situation again, this time from the state's perspective. States monitor traffic patterns and build roads where they seem inadequate for the number of cars and trucks on the road. States also subsidize bonds, economic development projects, and schools. This practice moves vast amounts of resources from city centers into sparsely populated areas to the detriment of the population centers.[68] Notice traffic patterns, schools, and economic development projects are an outcome of local land use decisions that have traditionally been made without regard to their impact on whole regions. Rather than purposeful planning that makes sense in light of the whole region, an after-the-fact issue is driving major decisions that further impact the land and its inhabitants.

As soon as the state decides to build a new freeway or road, pressure for development is put on the townships. They again

---

68 Michigan Land Use Institute, "Follow the Money: Uncovering and Reforming Michigan's Sprawl Subsidies," January 2005

compete against one another to attract strip malls, factories, and housing developments. And the cycle goes on. In Michigan, the cash-strapped state government can hardly function because so much of the budget goes to subsidize sprawling development[69]. The same is true of states across the nation. This engine of unplanned sprawl supports and is fueled by The American Dream.

The American Dream

The 1970's American Dream of a house and yard in the suburbs has evolved into the dream of the exurbs today because the 1970's suburbs have become part of the modern city center. These increasing circles and tendrils of "development" have led us to transform farm land into sprawled housing at an alarming rate. In the state of Michigan, on average land is being transformed from rural to urban eight times faster than the rate of population growth.[70] In the Northwest US, land is being transformed at more than two times the rate of population growth.[71] Similar statistics can be seen in the Pennsylvania Highlands and near most cities across the US. Where 50 years ago, there were discrete communities in townships, now it's impossible to tell where one community ends and the next begins. It's wall-to-wall strip malls and McMansions on fake lakes in the countryside and decaying urban blight in city cen-

---

69 ibid

70 The Orfield Report, "Michigan Metropatterns," April 2003

71 Theobald, David M. 2003. Defining and mapping Rural Sprawl: Examples from the Northwest US. Natural Resource Ecology Lab and Dept. of Recreation & Tourism, Colorado State University, Fort Collins, CO 80523 Retrieved 5/1/2009 from http://www.nrel.colostate.edu/~davet/Theobald_rural_sprawl-v1.pdf

ters. Those who are prosperous move out to the countryside, and the poor and minorities are left behind able to afford only the old, lead-paint-contaminated, energy-inefficient houses in the city centers.

Wealthy Americans

Let's look at the situation again, this time from the perspective of relatively wealthy Americans. In March 2007 *National Geographic* addressed the issue of urban sprawl in America using Orlando, Florida, as an example of *The Theme-Parking, Megachurching, Franchising, Exurbing, McMansioning of America*.[72] The Malvina Reynolds song about tract housing of the 1950-70s, "Little boxes on the hillside, Little boxes made of ticky tacky, Little boxes on the hillside, Little boxes all the same. There's a green one and a pink one, And a blue one and a yellow one, And they're all made out of ticky tacky, And they all look just the same" has been replaced by Walt Disney's message of self-fulfillment sung by Jiminy Cricket: "When you wish upon a star makes no difference who you are, anything your heart desires will come to you."

The new Magic Kingdom is exurbia in America, a place where vast tracts of manicured turf grass surround massive homes that have increased in size by an average of 63% over the past three decades, where restaurant chains create dining experiences, and where mega churches have sprung up to provide some semblance of community to the isolated people in this haven of plenty. Now Americans in exurbia leave home through their connected garage, get in their car, push the garage

---

72 T. D. Allman and David Brunett. "The Theme-Parking, Magachurching, Franchising, Exurbing, McMansioning of America: How Walt Disney Changed Everything," National Geographic, March 2007

door opener, and drive to their destination without ever making eye contact with their neighbors. The houses don't have porches, a feature that used to promote community. There's no need for a porch because the houses are air conditioned, so things like a cool evening breeze on the porch is a thing of the past. So is talking with the neighbors. To provide community, megachurches have evolved, huge wealthy entities that offer everything from parenting workshops, teenager entertainment, to support groups for divorcees. They are the Wal-Mart of churches; some even preach that it is a sign of God's approval to be wealthy! According to *National Geographic*, "The megachurch is the culmination, at least so far, of the integration of religious practice into the freeway-driven, market-savvy, franchise form of American life." Hey, why not? If you've industrialized your food, you might as well commercialize everything else into depersonalized, mega-consumption. And to make it all more convenient, there are drive in churches or you can even attend church by watching your favorite flavor of religion being sold on TV like a used car!

Notice what this does to individuals. Anyone who cannot drive is stranded in exurbia and dependent on those who can drive. Children and the old are prisoners of the non-community of dream homes. Everyone living there pays a constant tithe in windshield time getting to and from their job and activities. When gas prices go up, people have no choice whether to buy or not. They are locked into a tragic system of absolute reliance on import of food from far-flung places and import of energy to give them mobility. They are also locked into a system that isolates them from community, an important aspect of mental health for people. I regularly walk through Michigan versions of these neighborhoods. Frankly, I've never seen groups of children playing together outside. I suspect the neighborhoods are considered unsafe because people don't know their neighbors.

Poor Americans

There's another aspect to what these land use decisions do to individuals, this time the poor. Across the U.S., the poor are left mainly in the rotting city centers in old, energy-inefficient, toxin-infested housing and with poorly supported schools that are the discards of our parents and grandparents. In Michigan, inner city schools receive ~$2,000 less per pupil in statewide average funding than exurban schools. Similar stories on school funding can be seen in studies of major cities like Fort Worth, Texas, and New York[73]. Across the nation it is common to see the lack of resources and opportunities for residents in our rotting city centers that have not adopted green thinking.

There are no state-of-the-art grocery stores within walking distance of inner city homes, and the stores that are available frequently charge higher prices than exurban and suburban stores. Most jobs have moved out of the cities into the exurbs so the most entrepreneurial of the inner city poor pursue the most lucrative business opportunities available, mainly drug dealing and gang activity. Then, of course, we imprison high numbers of these people for their illegal activities at great cost. In Michigan, the state spends much more per year per prisoner than it spends per student per year. Even the lowest spending states on prisoners spend three or more times as much for prisoners than for educating children and the inner city poor frequently receives the least support of the students in the state.

---

73 USA, Government Accounting Office, 2002. "SCHOOL FINANCE Per-Pupil Spending Differences between Selected Inner City and Suburban Schools Varied by Metropolitan Area". GAO 03-024.

As a culture, we've left the poor with our discards and, because they lack mobility in the form of reliable personal transportation, we've let the "invisible hand of the market" raise prices, reduce educational opportunities, and separate people from employment opportunities for this audience held captive by their earning power.

According to the Government Accountability Office[74], inner city students in major cities they studied across the nation performed less well academically than their suburban counterparts. The gap in achievement is tied to factors like having a higher percentage of first year teachers, higher enrollments, fewer library resources, and less parental involvement. Education is the road out of poverty. How can our inner city poor escape from the trap they live in that was created by our land use practices and by the American Dream if our educational system reinforces their problems rather than solves them?

The results of this disconnected society can be measured by the number of people we now have in prison. A recent report from the PEW Center on the States found that one in 100 U.S. adults are now in prison, about eight times the incarceration rate of Great Britain. Higher rates for minorities are embedded in this figure—1/36 of adult Hispanic men, 1/15 adult black men, and an astonishing 1/9 black men ages 20-34. Spending on prisons has soared from about $10 billion in 1987 to over $45 billion now. We spend on average $1.17 per prisoner for each dollar spent on college education. Is this what we want? How can this be?

Thousands of "good" individual I, me, now decisions have led to disastrous consequences for the whole. It's a result of

---

74 ibid

our lack of significant cooperative social goals in land use and our lack of moral behavior as a culture and civilization. Land use practices drive social relationships so our current practices isolate people, damage communities, and impoverish us all. We have bought into rugged individualism and a selfish view of the American Dream that is depleting our quality of life as humans. We have allowed our governmental agencies to subsidize the selfish dreams of the wealthy to the detriment of our land, our children, and our cultural evolution. And the harm doesn't stop with humans.

Non-human Community

Let's take another look at this situation, this time from the perspective of the non-human community that shares the land with us. Creation is a naturally abundant system with all kinds of checks and balances that keep things livable for everybody. In an unaltered landscape, storm water systems evolve to capture precipitation as a valuable resource to take care of the needs of the inhabitants. This means that when it rains, little wetlands are formed where water infiltrates slowly into the underground aquifers and refills streams. All sorts of species are adapted to live within this system. For example, plants in drier areas have deep roots that channel water into the soil taking care of the water needs of the plants until the next rain and keeping the soil organisms supplied with water. And there are bugs, birds, bats, and beetles—always lots of beetles—that interact with one another, some eating the plants and others eating one another in a great system of checks and balances. These species depend on one another—insects, birds, and bats pollinate the plants so the plants can reproduce; many of the species rely on one another for a place to live, their ability to reproduce, and to raise their offspring. The American Dream leaves no room or resources for most of these species to live.

To build the American Dream, the shape of the land is altered, with the top soil bulldozed into piles where the soil organisms are exposed to conditions that kill them. Altering the shape of the land wipes out the little wet areas and changes the drainage patterns for rainwater so "stormwater" now becomes a problem that is taken care of by civil engineers through storm drains. After the land is shaped, it is covered with a thin layer of sod, a monocrop of a European plant, which is essentially an ecological desert for the species that used to live in that place. Turfgrass doesn't absorb rainwater well so irrigation systems are installed that draw water from nearby reservoirs and from underground aquifers; runoff is controlled by storm drains in shallow ditches. With the loss of the plant base of the food pyramid, the rest of the species are quickly lost. Those that aren't lost right away are taken care of by Scott's turfbuilder, weed killer, and insect control. Then when this bright green monocrop has the audacity to grow, the homeowner or their representative must keep it cut with a riding, air conditioned lawnmower that spews toxic hydrocarbons into the air and noise into the neighborhood.

The system's ability to maintain itself is lost and along with that is the loss of ecosystem services upon which people rely. The American Dream is an artificial utopia that relies on import of food, oxygen, water, material goods, and energy and export of mountains of garbage, waste water, toxic air emissions, CO_2 from respiration and energy use, and toxins from consumer products. The region loses any semblance of the biodiversity that brings stability to a complex adaptive system, so the reduced system can only persist in the absence of all but the most minor of perturbations. And each local patch taken over by the American Dream contributes one more loss that undermines the stability of earth's life support systems we have inherited from 4.56 billion years of earth's development. By rethinking our goals with a time, species, global perspective, we can prosper in the long term by learning from the lessons that can be read in earth's systems around us.

Pollinators

One ancient system most people are unaware of is the deep cooperative relationship between plants and the animals that pollinate them. We also seem to be unaware of how dependent we are upon this system. More than 87% of the world's flowering plants are dependent on pollinators for reproduction—pollinators move pollen from the male part of flowers to the female part of another flower. Seventy percent of all food plants rely on bees, moths, butterflies, beetles, and flies. Birds, mammals, and reptiles also pollinate plants in various ecosystems. Stephen Buchman, an entomologist at the University of Arizona, pointed out that one out of every three bites of food we eat comes to us courtesy of a pollinator.[75] Blueberry farmers consider bees to be flying $50 bills because of their impact on crop productivity. This ecological desert we are building in our cities, suburbs, and exurbs are unproductive, toxic places that don't support the pollinators. And our pesticide-laden industrial agricultural land doesn't support them either.

The loss of wild pollinators has been masked by businesses that keep honey bees and hire them out to farmers who need a

---

75 Cited from Jim Morrison, "How much is clean water worth?" National Wildlife, February/March 2005

crop pollinated, but in the past few years honey bee colonies have been collapsing. No one knows where they are going—there aren't a bunch of dead bee bodies around the hive. But they are simply gone[76]. One possible reason is a violation of one of nature's operational rules—nature thrives on diversity. Bee keepers take their hives to monocrops of almonds, fruit trees, blueberry bushes, and so on. Before the industrialization of our food system, bees (both wild and those cultivated by farmers) could visit a variety of food sources regularly. Now the only food available to bees at a given time is all one source—the crop their owner is being paid to pollinate. Imagine the problems of trying to get all of the nutrition your body needs from just one food source. If nothing else, I personally would leave the "hive" and never show up again. The local rule has also been violated with bees. Bees are brought from all over the U.S. to pollinate particular crops, making it so diseases can spread quickly. In the case of bee colony collapse, a viral disease (or possibly several viruses together) is implicated although other factors are almost certainly involved. One way or the other, I hope the implications are clear to you. Now it isn't simply a matter of missing butterflies from our gardens, it is becoming a matter of missing fruits and vegetables from our tables. To my way of thinking, this is enough doom and gloom for the time being. Let's take a look at how we can turn our private land use from the practices of Mordor into those of the Shire.

Changing Our Land Use Practices

Far-thinking people in communities around the world have found effective ways to control development so it is a net

---

[76] Erik Stokstad, "The Case of the Empty Hives," Science, 18 May 2007, pgs 970-972

positive rather than a scourge. I've flown over France a number of times on my travels to Madagascar. When you look down on the French landscape, you do not see houses dotted all over the place like in the U.S. Instead, there are compact villages and cities surrounded by fields, fence rows, and forest patches. The same is true of parts of Kentucky and Oregon due to some forward-thinking state laws enacted over the past 30 years. If state laws legislate large parcel size in the countryside and don't limit boundaries of urban growth, sprawling development is forced to occur by law and promoted through economic forces like the price of land. On the other hand, by having sensible state laws, local planning commissions must function within a framework that makes sense for people and the non-human inhabitants of the land. Agricultural land can be preserved and taxed at a rate appropriate to the use of the land rather than taxed out of existence by the proximity of neighboring housing developments.

State and federal government has a definite role to play in the solutions to the sprawl problem. Tax dollars should not follow growth; instead tax dollars need to be the impetus to a wise outcome for our civilization. They can be used as a driver to promote sustainable social investments in transportation infrastructure, parks, greenspace, preservation of farm land, and the water and sewer systems so important to disease control. Imagine living in a country where sprawl would be forced to pay for itself. The developers and their customers would have to pay for lost local food production, loss of ecosystem services ($50/bee), future costs of EPA-approved sewerage treatment plants, future costs of water treatment facilities, improved roads, larger or new schools, more police and fire personnel, the cost of their neighbors having to re-drill their wells if they have put a well into an aquifer, and so on. Suddenly the economic equation that promotes sprawl would be tipped in favor of redeveloping the trashed land in city centers. This means our development must be more compact.

Compact Development

If you are like most Americans, the words compact development and the idea of living in crowded conditions with your neighbor's windows aimed directly at yours has to raise goose bumps up your back. It does mine. But this is where William McDonough's design principles become important. Remember he said "Design is the first signal of human intention." If our goal in compact development is to cultivate a wide variety of locally adapted communities with a diversity (remember nature's organizational rules) of choices for people, then we can have our cake and eat it too. For people who live primarily inside their houses and keep their yards only as a showplace setting for their home, houses could be arranged with private views and a cooperatively maintained showcase setting, cared for either by professionals or by the residents depending on the desires of the people involved. Or possibly a mixture of people who like yard work and those who don't? Or a mixture of other amenities? Thousands of community designs are possible that meet the particular needs of the people involved if we signal our intentions accurately.

There's a useful book called "Getting To Yes: Negotiating Agreement Without Giving In"[77] that has been reprinted numerous times. One story in the book is instructive here. Three

---

77 Fisher, R, W. L. Ury, and B. Patton. 1981. "Getting To Yes: Negotiating Agreement Without Giving In". Penguin Press.

children are arguing over who gets an orange. Traditional thinking would be to cut it into three pieces and be done with the argument. None of the participants would get everything they want though. By negotiating and communicating clearly however, the book points out how each participant might get 100% of what they want. Possibly one wants the orange for its peel for an ingredient in a cake, a second wants the fruit to eat, and the third might want the seeds to plant for a science experiment. As we think about concepts like compact development, we should be using this sort of thinking. Most Americans own a sprawl home for only a few aspects of that home. Clever, sophisticated compact development designs could be developed to provide each person in the development with 100% of what they value.

In the process, we could take a big step toward solving many of our social, economic, and environmental problems. Global warming, traffic-related fatalities, social isolation, obesity, energy costs, loss of biodiversity and farmland, and many other interdependent problems could be addressed simultaneously if we were to insist on change in our land use laws. For example, effective designs would allow us to develop efficient mass transit that would improve our mobility, decrease our windshield time, decrease traffic-related health impacts, clear our air, save energy, decrease global warming, and increase space for wild pollinators. Also, by freeing up land in the region of our homes, it would be possible to eat a locally adapted diet rather than an industrial food system diet of corn-based pseudo food that is not good for our health. One group we must pay particular attention to as we think about redesigning our land use policies is farmers whose families have owned their land for several generations.

Farmers as an Endangered Species

From the perspective of traditional farmers, changes like the ones proposed above are a bad deal. Traditionally U.S. farming doesn't pay high wages and the work is hard. Farmers farm because they love the land, but they get old just like the rest of us. Most children of farmers aren't farmers themselves. This, coupled with the sprawl system we've allowed to develop in the U.S., has created a situation where farmers can retire wealthy by selling their land to developers. The big old tree on their property can bring in enough money to send one of their kids to college and their wood lot can be the farmer's insurance against a bad crop year, while city dwellers depend upon it for esthetic value and also for keeping gases in the atmosphere balanced. Notice the urban dwellers depend upon the farmer's land but don't help pay for it; they treat the privately owned farmland like a commons. As a civilization, we must address the perspective of the farmers in an effective way if we are to repair our system of private land use.

Primary issues include methods for farmers to get the value out of their land without promoting sprawl. We need to develop systems to recruit each new generation of farmers who consider stewardship of the land and its productivity to be their life's work. The fate of wood lots and big old trees can't be the sole choice of the landowner but we cannot expect that maintenance of these vital pieces of ecosystem integrity should be provided to the commons without financial gain for the traditional owners. Some systems already in use in small patches of the U.S. include community supported agriculture (CSA), purchase of development rights, and federal land bank programs, all of which are effective within the constraints of the sparse money available. For example, the population of farmers in rural Missouri, most of which has not experienced sprawl, has aged over the past 30 years and now many earn their retirement income from the CRP (Crop Reduction Program) that pays them to not grow crops. CSA is an inventive system whereby houses surrounding isolated remnants of farm land pay a yearly fee for a share of the fresh produce grown on the farm. Everybody wins with this system: the farmer has a guaranteed income and the people surrounding the farm have a reliable source of fresh, local produce and meat. Here again, nature's operational rules of diversity and local adaptation should be our guideline. Imagine how much money would be available to ensure a steady supply of local food if our governmental entities quit spending to promote sprawl and spent a portion of that money to preserve the art of farming! Changing our federal and state goals from subsidizing the desires of individuals to the detriment of the land and our entire population to goals that recognize the necessity of living within the rules of nature would be a major step forward in bringing the pollinators back while keeping family farms in business. Both are important in delivering each bite of nutritious, esthetically pleasing food to our mouths.

Re-knitting Nature around Us

Another set of concerns and solutions come from the community level. Gray infrastructure—the system of roads, sewers, schools, and stores needed to support a human population in any area—is built automatically with development of the land for human purposes. Green infrastructure needs to be protected and built at the same time—this is an interconnected system of intact patches of the various biological communities of an area connected by corridors of natural vegetation that act as roads for non-human species to move from place to place. The important issue for biological communities is comparable to the type of connectivity you feel when you're in your car—there's a system of roads, parking lots, and garages that make it easy for you to get around. We must provide the same type of system for the species around us so they can live and reproduce. For example, from the perspective of a frog, a road is more of a barrier to movement than miles of river. If frogs can't move, then local extinctions of populations become permanent. It doesn't take too much of that and soon there are no frogs in the region. That's a major loss because frogs are great bug eaters.

Green infrastructure can perform all sorts of services for the human community and save money at the same time. New York City has historically gotten clean water by protecting the forests and wetlands upstream from the city. Instead of building multi-billion dollar water treatment plants, they have let nature provide clean water for them at a fraction of the cost. Property values are higher when they abut green space or even tiny streams. Having nature nearby provides recreational opportunities, higher land values, better mental health, abundant pollinators, and nearby sources of food. We all win when we purposefully protect the lives and livelihoods of our non-human neighbors. This cannot stop at the outer edges of city centers.

We would benefit by rebuilding our city centers into a system of small communities of mixed-income housing instead of rich and poor neighborhoods. Each community would benefit by including native mixed-species urban forests, numerous rain gardens to act as retention ponds for rainwater and as islands of native vegetation to house frogs and pollinators, and green roofs with deep soil that reduce heat island effects and increase building insulation while providing habitat for birds and pollinators. We have to get rid of the ugly pink mulch next to our houses and at the base of trees—this stuff decreases the amount of viable habitat for toads and the species toads use for food. If we value nature and nature's services, we in urban areas need to value and promote nature near us just like our rural counterparts. In other words, building green infrastructure means re-knitting the fabric of nature into all the spots we live.

If this is going to happen, some local laws will have to change. Weedy lawns are now against the law in many places. Even native flower gardens are sometimes illegal. Going native doesn't mean things have to be all brown and ugly. As a matter of fact, well designed collections of native trees, shrubs, perennials, and annuals flower longer and are more drought-resistant that the finicky exotic ornamentals now frequently planted. They also are resistant to disease and do not need to be fertilized so we can accomplish several goals all at the same time—reduce toxins in our environment, protect ecosystem services, and provide places where we can safely play outside with our children. I hope you see the correct goal we must have for nature is not just preserving the status quo. We must also preserve nature's ability to evolve to meet changing conditions. If we can accomplish that, then we will also have preserved the ability of our children's children to live here.

Re-knitting Human Community

There's one last piece we should talk about before closing this chapter. That's the idea of human community. In my mind, the saddest victims of The American Dream are our children and our senior citizens. Let's put age before beauty and talk about the old people first.

I've recently retired and I've found the social expectation is for retired people to spend their time relaxing and finding things to do to keep them pleasantly entertained. Younger retirees buy RVs that get a few miles to a gallon of gas to travel all over the U.S. stopping here and there to while away some time. As retirees get older they move from independence to a monocrops of assisted living facilities where they spend more and more time watching TV in their isolated islands of old people. I once took my mother and her 96-year-old friend, Margerite, to a movie. When we got to the theater, Margerite commented as she came in the door, "Oh, this is where all the young people are!" Segregation of people by age groups isn't our best idea. And the idea that at some age and/or wealth level, people have the right to essentially opt out of a productive role in our civilization is plain wrong.

It is especially wrong when the expense of supporting this waste of human talent is more than the civilization can bear. As our population ages, we hear dire warnings about how few workers there will be to support an increasing number of people on Social Security, a situation that would not exist if older people maintained some level of productivity throughout their lives. Maintaining engagement in productivity would make it so the young would not grow up in a world where old people are warehoused out-of-sight and out-of-mind. Most of us will get old. It seems to me it is worthwhile to expect the old to contribute to society at some level for their entire lives. And it seems to me that we are losing a great deal if we ignore the productivity that is possible from the activities of those who no longer work at regular jobs. The point is that society really ought to expect some return on its investment in Social Security and Medicare. Somehow we must reintegrate our older citizens back into useful roles in our civilization and part of that role should involve contact with our young. Both groups would benefit from the interaction—the old would have to become less isolated and self-absorbed while the young would gain wisdom from an understanding of the inevitable progression of life.

Children have been the other victims of our current American Dream. It may be the great error of our time to attempt to form monocultures of our kids in huge schools in the same way we produced automobiles and corn. Aggressive behavior increases in single-age groups of children and the real-life ability to interact with people of different ages, perspectives, and backgrounds is not developed well in this setting. According to a report in *Parade* magazine (March 2009)[78], children learn better when grouped by what they already know rather than by their age level. A successful example is from the Chugach School District in Alaska that adopted multi-age groupings of students in the 1990s. After five years, student achievement scores had jumped from the bottom quartile to the top quartile. A few other school districts around the nation are beginning to adopt this system. Think about it. Rather than having fifth graders in special education because they can only read at the third grade level, these students could be in a classroom with other students who are at the same reading level. A 12-year-old who is ready to start studying calculus could do that rather than spending six more years being held back to the level of most students their age. Think about how much money we would save from the expense of special education alone! Think about the self-image of our children as they no longer would have to be below average or targeted as learning disabled. Think about how more effective our teachers could be if they were working

---

78 Retrieved May 1, 2009. http://www.parade.com/news/intelligence-report/archive/the-end-of-grade-levels.html

with a group of students all of whom had the same foundational knowledge.

Instead of huge industrial schools, we would be much better off with well designed, intensive learning situations where the teachers are able to get to know their students individually. Currently teachers are moved around school districts like pawns on a chessboard to meet the needs of the school district. Many are not permitted to hone and develop their skills in a stable setting with students they know personally. In our deep history, our children learned all the skills they needed to be successful by learning from the adults around them. They didn't get a new teacher each year and weren't treated like a corn crop, all growing, developing, and being harvested at the same time.

The anger and frustration of our children in this industrial educational system is apparent. The problems we currently have with bullying, gun violence, and absenteeism would certainly be reduced if the learning environment more closely resembled the learning environment of ancestral humans. Small, local schools that team in broad networks for special services would allow kids to walk to school rather than ride a bus an hour or two a day. In small communities embedded in city centers, it would also be possible to protect our children, possibly by a system of senior citizen watchers, so our children would be free to play outside. If you think about it for a minute, you realize children are a "small, local" part of our civilization. Their nurturance should be locally adapted, just as nature's solutions are locally adapted.

We've spent a good number of years with subconscious, poorly-designed goals in our relationship with the land. That relationship has translated into unintended consequences for ourselves and the non-human species around us. We're all suffering and our culture in collapsing under the weight of the destructive, self-serving goals of the American Dream. If we are thinking green and making choices based on a time, species, global perspective, we must purposefully design a relationship with nature that respects everyone living on the land and making a living off of it.

Important Concepts of this Chapter

Land use traditions, a legacy of America's unique history, control the structure and function of our modern landscape. This heritage also endows us with the national parks as well as with the system of public lands we enjoy, all of which are endangered as over-exploited commons areas.

Americans have exceptional rights associated with private land, where the landowner has almost complete control over everything, even when the landowner's activities harm others.

Traditionally, land use in America is controlled at the local level with few restrictions from the state and federal government. Millions of I, me, now local decisions of small local governmental units on land use have resulted in disastrous land use practices in the U.S.

The American Dream of a home in the country has evolved into a disorganized, sprawling exurbia where the costs of providing roads and services far outweigh the local tax dollars attracted by the development.

Much of the American middle class is widely dispersed in housing subdivisions built on former farmland in ever-larger houses that can only be accessed by motorized vehicle; this dispersed community structure is especially damaging to children and the old.

The poor are concentrated in the cast-off, lead-paint-contaminated housing in city centers with educational opportunities limited by shrinking school districts, old buildings, and shortages of employment opportunities for the parents.

Nature's life support systems, including the pollinators essential to food crop productivity, are systematically dismantled in our drive to sanitize our living spaces into vast showcases of turf grass and ornamental plantings.

Food for the future depends on social arrangements that value farmers, farmland, and farming as a viable profession for young people.

Urban people cannot expect rural people to provide intact ecosystem services alone. Cities must become places where nature is re-knit into the fabric of communities in the form of green infrastructure—urban mixed-species forests, rain gardens as small wetlands of native species, and green roofs.

The goal of our relationship with nature should be to preserve nature's ability to continue evolving, not to preserve a status quo of a by-gone time.

Land use and our ability to form viable communities are inextricably intertwined. In the design of our communities, we must pay particular attention to our youth and to our older citizens. Warehousing old people and isolating our children in mega-sized "monocrop" schools does not allow us to form communities as is natural for humans. Our children are our future so they should be nurtured in small schools that are safe and locally adapted.

* * *

The final challenge I will cover deals with how we humans go about conflict resolution. War is an accepted part of our activities, the education of our children, and our ethical rules. During war, just about anything is acceptable, including the killing of non-combatants as "collateral damage." Whole economies today are built on war. But there is a wealth of alternative strategies to be learned from our fellow species in this ancient place. Possibly there are better ways to resolve conflict?

Chapter 9

War—An Evolved Piece of Cultural Insanity

So far this book has mainly addressed U.S. culture, its problems from a time-species-global perspective, and some solutions that are already being implemented in a local, piecemeal way. This chapter is about war, a phenomenon accepted in nearly all human cultures without much question. Frankly, it seems war is such an integral part of the human experience that we accept it as a given alongside birth and death. And many of us, even those who have lost relatives in war, consider it to be a necessary and desirable thing. We honor war heroes and those who have lost their lives "defending their country" even if we happen to be the aggressors in the conflict. Because this planet has become ever more crowded and because our war activities are so damaging to people and the earth, it is critical we examine our assumptions about war and develop a time, species, global perspective on it. If we are thinking green with this perspective, it becomes apparent our current global practices for conflict resolution are unethical.

History of War

I remember studying history during my K-16 education, a subject I really didn't enjoy much. Hindsight has helped me understand why because now I am intrigued by human history, the archeological record, the fossil record of life on earth, and the history of the universe as much as we can understand it.

The history I studied as a child was a lexicon of kings, wars, and migrations of select groups of people. Always the lists started at some arbitrary time and unfolded toward the present. As I look back on that impoverished story now, I know why it wasn't interesting—things of substance that made a difference in the lives of women and children weren't covered—things like when certain foods became available and why particular social rules developed and changes in the practice of killing children. The story told to us in the guise of history was the story of powerful men and a few women who lived in a world of wars and conquest and entitlement, almost like a child's dot-to-dot puzzle with no lines drawn in for common people.

Now I must ask myself why rulers and wars are considered so important that children must memorize this list of dates and names. Why not teach what life was like for common people at different stages of our history and in different cultures around the world? That seems to me to be much more important information for children to study, information that would have helped us become citizens of the world rather than soldiers in our country's army under the command of generals in service of agendas hidden behind war slogans. The key question here is why humans are so war-like. It seems a species as intelligent as us would have come up with better ways to solve and avoid serious conflicts.

War Is a Uniquely Human Practice

All species must resolve conflicts but it is rare for species to conduct wars to do it. Among mammals, the latter practice seems to be restricted to humans and their nearest biological relative, the chimpanzees. Most vertebrate animals engage in conflict, mainly originating in the competition between males for defense of mating territories and defense of the right to have sexual access to females. Other conflicts occur over food or nesting/denning sites, but those conflicts are brief and relatively minor. The severity of the sexual conflict makes sense from an evolutionary perspective because if a male doesn't get his sperm into a fertile female, he is total loser in the struggle for existence. As a matter of fact, from a male's perspective, the more females he can fertilize, the more he is favored in the evolutionary race. That's an important enough reason to get into some really violent conflicts because in an evolutionary sense, it's mate or die genetically, so there's a lot at stake. But even in the face of such strong forces, war, especially the scorched earth wars of humans, haven't developed. I'm not suggesting that all species other than chimpanzees and humans are pacifists—they can, in fact, have some royal battles.

About the time I was finishing my doctorate, I was teaching on a temporary basis at Albion College, a small private college in southern Michigan. One of the courses I was teaching was Vertebrate Zoology and, as a new professor with a small child, I was struggling to provide adequate academic content with my limited time for preparation. Using the adage that one picture is worth a thousand words, I took the students over to my rental home one day to show them my lizards. At the time, I had three—two males and one female. You probably have a vision of a lizard in your mind—small, brown, long tail, fast. That's a wrong vision of these lizards. They were mature green iguanas, each 4-6' long and weighing 8-12 lbs apiece, and fairly docile most of the time. In the wild, iguanas are territorial. Iguana males stake out their turf and guard it while females move freely between territories choosing to mate with some males and not others on the basis of criteria I have never understood.

Anyway, I figured I had all the important elements to teach these students about lizards and lizard behavior. Great plan, or so it seemed at the time. Everything started out nicely. There were 20-25 senior level students in my living room and I brought in the larger male and the female, Trav and Piggy, respectively. Trav immediately did his male lizard thing and bobbed his head up and down in the predictable iguana pattern of challenge to any unseen iguana males in the area. Piggy just ignored him as is characteristic of female iguanas—maybe we human females should learn some lessons from them. So that was all well and good, but not very exciting for the students. Reptiles aren't like mammals—they sit really still rather than twitch constantly—so soon the demonstration was becoming boring for the students. To demonstrate the male territorial behavior further, I brought out a mirror for Trav. He quickly flipped out his gular flap, a web of skin below his chin that is used as a mechanism for collecting solar heat and also used to make the male look bigger in a territorial fight. He also did more head bobbing. That was kind of exciting, but still not all that great for the students.

Finally I brought out Skinny Butt, the other male iguana, and as his name implies he was a 90 lb weakling compared to Trav. Still they were the same age so it wasn't like letting a big bully knock a little guy around. In short order, the two males had flattened their bodies vertically and turned broadside to one another with gular flaps out and legs extended. They were each making themselves look as big as possible. Then they danced in a circle bobbing their heads and wiggling their tails at one another—the messages were clear—this is how big I am, this is how much strength I have, and I will beat the crap out of you if you don't get out of here right now. Neither male left so the fight escalated into some tail-whipping and ritual bites of the opponent's elbows and knees.

You can tell male iguanas from females because the males have big jaw muscles that are used only in their male-male ritual bites. So the elbow and knee nips are a way of communicating to their opponent how strong they are and how much damage they could do if they are pushed further. Well, about the time Trav gave Skinny Butt a tail whip and a nip on his knee, Skinny Butt jumped like a flash right over the couch and into the midst of a bunch of students. Lizards move much faster than mammals. The students were jumping over the couch and running out of the room trying to get away from the lizard fight. Fortunately no one, lizard or human, was hurt in the fray. That's the one and only time I allowed those males to be in a room together. Still, there are some lessons to be learned about conflict resolution from this lizard battle.

As occurs in most species, conflicts are resolved with ritual fights, not fights to the death. Violent demonstrations of strength and power are shown to the enemy, but animals rarely fight to the death. According to Richard Wrangham[79], of the nearly 5,000 species of mammals only humans and our closest relatives, chimpanzees, live in social groups with their relatives and cooperatively make and carry out plans to kill neighboring individuals or groups of their own species, especially when those groups include females and young. Primate species that split off early from the chimp-human ancestor don't seem to make war on their own species, but we do. Why is that? Why haven't we developed ritualized fighting between groups to solve

---

79 Richard Wrangham, "Demonic Males: Apes and the Origin of Human Violence," 1996, Houghtin Mifflin

our conflicts rather than scorched earth methods like atomic bombs? One answer to these questions can be seen in the feeding ecology of our ancient ancestors.

The Role of Feeding Ecology

Feeding ecology deals with the distribution and abundance of food, a key factor that influences social relationships among females. If the primatologists like Richard Wrangham of Harvard University are right, this is a woman's issue. If food sources are scattered throughout the habitat rather than bunched in trees a great distance apart, then females can stick together and form coalitions with one another that hinder male dominance and violence. If the distribution of vital food sources forces females to forage independently in order to find enough nutrition for themselves and their young, then isolated females are vulnerable to the demands of males as they vie with one another for access to and dominance over the reproductive potential of the females. In this setting, rape and infanticide are commonplace; males kill the young of other males so the female can produce an offspring sired by the new male. Essentially, violence and competition arise out of isolation of females caused by the distribution of their essential food sources.

Closely related to the chimpanzee-human line of evolution are the pigmy chimpanzees or bonobos. Bonobos live in larger social groupings than chimpanzees in an area of Africa that did not have gorillas on the ground during a critical period of the evolution of the species. Without gorillas bonobos had access to a fairly continuous food supply in the landscape so the females evolved the habit of staying close to one another and forming strong coalitions. To offset the aggressive energy of males, bonobos have developed a "make love, not war" set of behaviors. Given popular images of males, I would think human males would much prefer a bonobo system to a chimpanzee system, but I guess we can't choose our ancestors. So 90-95% of human cultures are war-like and have patriarchal governance structures—most presidents, dictators, and monarchs are male and males are the sex most likely to attack, conquer, and kill their competitors. As we've evolved during the past five million years, humans have escalated our war-like tendencies and war materials along with the development of our technology and economy.

Predicting the Winners of War

Jared Diamond has written an insightful book on this topic called *Guns, Germs, and Steel: The Fates of Human Societies*[80]. Diamond uses a global view of the past 13,000 years to explain his thesis, the period since humans started settling down in villages. By 13,000 years ago, humans were pretty well spread around most of the continents and had diverged into locally adapted cultures. The question Jared Diamond asks in his book is "What determines the winner of conflict when two cultures come in contact?" Notice war-like behavior is so deeply engrained in our species that there isn't a question of whether there will be conflict; it's a question of which culture will win. Diamond's answer flies in the face of the common assumptions we make and earned him a Pulitzer Prize.

Given classic U.S. history training that is prevalent in American schools, the automatic subconscious answer is always based on the assumption that white European descendents

---

[80] Jared Diamond, "Guns, Germs, and Steel: The Fates of Human Societies," 1997, W.W. Norton & Company

will win because white people are inherently smarter and more industrious than darker-skinned cultures. Contrary to popular assumptions, Diamond presents a plethora of data that shows race is not a determinant of the outcome. To understand Diamond's ideas, it's important to think of the various original cultures around the world as local adaptations to the particular environments available. For example, Inuit culture is adapted to resources and climate within the Arctic Circle and the various versions of Chinese culture are adapted to the resources and climates of Asia, and so on. When these well-developed cultures come in contact with one another, they differ in the weapons they have developed (or not developed), the material they use for various purposes (e.g. steel is much more durable than stone tools), and the germs (diseases) to which the population has developed immunity. The historic feeding ecology of each culture has a huge impact on the development of these characteristics and in the cultural mores.

Jared Diamond tells the story of two cultures on the Chatham Islands located ~500 miles east of New Zealand. Both cultures derived from the same ancestral Polynesians who colonized the area around A.D. 1000. Because the southern island was too cold to raise the Polynesian crops, the southern people developed a hunter-gatherer culture where crop surpluses were not available to support non-hunting specialists who would have had time to develop increasingly complex technology, including weapons. Without farming and without nearby islands to colonize, the carrying capacity (number of people who could be sustained) of the island was limited. Diamond says of them, "With no other accessible islands to colonize, the Moriori had to remain in the Chathams, and to learn how to get along with each other. They did so by renouncing war, and they reduced potential conflicts from overpopulation by castrating some male infants. The result was a small, unwarlike population with simple technology and weapons, and without strong leadership or organization."

The northern, warmer island was different. There the climate was suitable for Polynesian agriculture and the land masses involved were larger. Population size grew quickly and periodically densely populated areas engaged in fierce warfare with neighboring groups. The food surpluses from their agriculture allowed them to support trade specialists and even part-time soldiers. The selective pressures on the culture of these northern Maori consisted of an obligatory arms race in self-protection and armed conflict. The dichotomy between the warlike Maori and the peaceful southern Moriori came to a disastrous slaughter between November and December of 1835.

First an armed party of 500 northern islanders arrived at the peaceful southern island and then 2 ½ weeks later a second party of 400 more northerners arrived. The first to arrive walked through the Moriori villages announcing that they were conquered and were now slaves—they killed anyone who showed resistance. The Moriori could have fought off the first invasion because they outnumbered the invaders 4:1 but they had a tradition of peaceful settlement of disputes. In a council meeting they decided to offer peace, friendship, and a division of resources to the invaders. That sure didn't work.

Before they could even deliver their offer, a second group of Maori arrived and within a few days they had killed, cooked, and eaten a large portion of the Moriori population. Running and hiding didn't help—all were attacked—women, children, and babies.

This story has been repeated all over the globe through the centuries—well, maybe not the cooking and eating part. The most poignant part of it is the idea that any culture that developed along peaceful lines was eventually overtaken by a more war-like culture in search of additional resources. We pay a heavy price for the frozen accident of feeding ecology that caused the chimpanzee-human common ancestor to "make war, not love."

We Live with Our Frozen Accident

Of course, the number and types of human cultures we have today have become ever more complex. Traditional local wars still occur with regularity. For example, the recent slaughters in various parts of Africa and the ongoing conflicts in the Middle East are noteworthy. In addition to local wars, we've developed some more insidious versions of war. With globalization and our ability to communicate almost instantaneously, relatively small groups of war-like people can connect with one another and wage diffuse wars on the rest of humanity in service of various ideologies. I'm sure you think I'm referring only to Al Qaeda but I'm also including powerful transnational businessmen who manipulate dictators in service of economic and/or political ideologies with the purpose of making huge profits by gaining access to resources sited in various nations. For example, the clash between the economic policies derived from the theories of John Maynard Keynes, engineer of the U.S.'s New Deal policies, and Milton Friedman, Nobel Prize winner in economics who championed free markets almost as a religious ideology, caused blood baths in Latin America during the latter half of the 20th century[81]. The U.S. is central to much of this sort of conflict because of our economic power.

With the collapse of the Soviet Union, the U.S. has become the global leader in military might and is the home base for ideologies that escalate mechanisms designed to extract resources out of the homelands of other cultures—at least for now. China, along with its economic growth, is putting more and more resources into building its military capabilities. Those who are not concerned about the shadow of U.S. military and economic might hanging over the heads of cultures around the world should become concerned at the specter of Chinese supremacy in that role. Now when it is abundantly clear that we may be losers as well as winners in the military might arms race, possibly we might consider changing our cultural behavior so we do not stimulate other nations in the world to develop the horrible weapons of war we have pioneered? Fortunately, this ancient place has developed many models we might adopt.

Conflict Resolution Nature's Way

Arms races are a natural outcome in complex adaptive systems when replicating/mimicking elements engage in fairly balanced competition with one another (Chapter 3). This was true throughout the cold war between the U.S. and the U.S.S.R. The same thing occurs in non-human systems. For example, forests of all types tend to have their tallest trees all about the same height. The tree species evolved and counter-evolved to gain maximum light, a key resource trees need to grow. The only counter example I've ever seen to this forest phenomenon

---

[81] Naomi Klein, "The Shock doctrine: The Rise of Disaster Capitalism," 2007, Metropolitan Books, Henry Holt and Company, LLC

is along the west coast of Madagascar where the baobab trees live. These trees have massive, fat trunks, some of which are 30 feet or more in diameter. They tower over the surrounding trees that are no more than about 15 feet high. I suspect the baobabs avoided an evolutionary arms race by restricting their limb growth to a network of tiny branches at the top of their massive truck; these trees quickly grow leaves and flower, and then the leaves die; essentially baobabs have restrained themselves in such a way as to avoid shading their competitive vegetation— the shorter members of the baobab forest have abundant branch and leaf growth. I think the moral of the story here is that U.S. military might and protective behavior does not have to be done in a way that stimulates the development or growth of an arms race with other nations on this small planet. Strength coupled with non-aggressive, respectful behavior would have a very different impact on other countries than strength coupled with imperialism.

Another example of nature's solution to aggression can be seen in Eastern Phoebes and other species of small birds common where I live. Among other species, Phoebes are parasitized by cowbirds, which lay their eggs in host species' nests. When the eggs hatch, the cowbird hatchling can kill the host nestlings but cowbirds leave the host nestlings alive because they get more resources from the host parents if the host's young are present[82]. For years ornithologists thought that the host birds didn't know the difference between their own and the parasitic eggs or offspring, assuming the birds were too dumb to tell the difference between their own and the parasites. It turns out

---

82 Rebecca M. Kilner, Joah R. Madden, and Mark E. Hauber, "Brood Parasitic Cowbird Nestlings Use Host Young to Procure Resources." Science, 6 August 2004, pg 877.

that they do know the difference—the cowbirds keep watch on the situation, and if the hosts kill the parasitic young, then the reproductive success rate of all the birds in the area decreases because the cowbirds raid all the nests and kill the young. Even birds are subjected to the Mafia but they have developed ways to live peacefully with something even as intrusive as brood parasitism. This is an example of social cooperation among species to protect the quality of their nesting neighborhood, a commons they all rely upon. Yes, the cowbirds are permanent cheaters in this commons, but because it's impossible for the birds to cause the extinction of the cheaters, they all have the best reproductive success by their current social arrangement. Why can't we do the same? It may cost us a bit to put up with some threat and unfairness plus we may have to come up with some definite social arrangements to control the severity of the problem, but the alternative, war, is much more damaging to us and to our ability to persist in the long term on this earth.

The human population has grown too large and the earth has gotten too small for us to cling to this ancient, barbaric behavior. From the standpoint of the environment as well as our social, cultural, and economic evolution, we must outgrow the destruction associated with war if we are to survive on this planet. We think of war in terms of economic costs and in terms of the lives lost but possibly an even longer-term impact is on the ability of earth's life support systems to sustain our children and ourselves, because the environmental impacts of war and war preparations are devastating.

Environmental Impacts of War

Most of us are aware of war impacts like the potential of "nuclear winter," the devastation of Nagasaki and Hiroshima, the destruction of lives from land mines, the air and health impacts of the burning oil wells from the Gulf War, and the shattering aftermath of the bombing raids of WWII. But few of us are aware of the ongoing severe environmental damage caused by war preparations. For example, Rocky Flats near Denver was a weapons development facility during the Cold War—DOW Chemical and the Atomic Energy Commission were involved there. One of my former students, Mark, was working for the Colorado Division of Wildlife; he said he had been to ponds at Rocky Flats where migrating ducks land for rest but promptly die and sink. This 30,000+ acre site is still being monitored and cleaned up as much as possible by the Department of Energy. There are also ongoing claims through the Employees Occupational Illness Compensation Program associated with human exposure at Rocky Flats that may cost even more money in the future.

All over the U.S. and around the world there are massive tracts of land contaminated with nuclear waste, unexploded ordnance, and deadly chemicals that impose yet another massive cost to the Earth from our war activities. We put up hazardous materials signs to keep people away but the wildlife can't read; radioactive ducks and deer move freely across the land. Many of the materials on these sites are so hazardous they will last longer than any written language. How will the people of the future know to stay away? Is this the heritage we wish to leave to our descendents? War seems to be so deeply engrained in our culture that an identified enemy isn't necessary. Even with the end of the Cold War, things don't seem to be winding down and the environmental problems keep escalating.

Back in 2003, the U.S. Senate Committee on Environment and Public Works held a hearing on the "Impact of Military

Training on the Environment." The question was whether the U.S. military should be exempt from the Endangered Species Act, the Migratory Bird Treaty Act, the Clean Air Act, the Marine Mammal Protection Act, as well as a variety of federal toxic waste laws. To put things in perspective, the U.S. military controls 25 million acres, approximately 1.1% of the total land area of the U.S. Originally those lands were "in the middle of nowhere" but now are surrounded by housing, strip malls, and interstate highways. Because of this incompatible, myopic land use planning adjacent to military lands, DOD land has become the last refuge of over 300 threatened and endangered species who have fled adjoining development, and the military has become increasingly constrained in running "readiness training" for their personnel.

Readiness training consists of shooting guns, blowing up bombs, running tanks over the land, running bombing missions from low-flying planes, digging fox holes, and a variety of other activities that don't lend themselves well to the values of conservation and preservation of the species living on these vast tracts of land. The problems don't stop with terrestrial species because the long reach of the U.S. military extends into space and under ocean surfaces around the globe.

The military now uses sonar off U.S. shores and around their ships in foreign seas to detect threats. From the perspective of marine mammals who use echolocation to communicate with one another this invasion of water with sonar is comparable to blasting rock music at the highest volume possible. There have been numerous reports of beached whales after military sonar use—the whales apparently are driven out of the water with ears bleeding. I feel like that sometimes when vehicles roar past my house with their amplified bass music at full blast, talk about an intrusive, traveling menace. That's what our military sonar use in the oceans is: an intrusive menace, but to what ends? The military is still out there in the ocean behaving as though there are hidden Soviet submarines sneaking up on us to launch nuclear warheads at our major cities. Meanwhile, the main invasion we are experiencing is a quiet flood of poor, illegal immigrants from the south who are fleeing from their home country in search of food and employment.

Only the Big Have a Say

Possibly I am naïve about war, but it seems to me that globally, this behavior is a complex adaptive system subject to the selective forces we place on it. Small countries have little say in the situation and must do their best to survive. Using our outsized military might to "spread freedom" —a euphemism for spreading free market economy—is an oxymoron. Trade can occur only between equal partners, both of whom gain desired products or materials in the exchange. Anything else is an exploitation based on military and economic strength. Large countries like our own bear a heavy moral responsibility to seek ways to avoid stimulating arms races and to avoid exploitation of weaker countries. In this context, at some level Al Qaeda is the desperate fight of a small group of extremists who have redefined the rules of war so they can fight to protect their way of life from Western incursions of economic and military might. It was no accident they attacked the World Trade Center, the symbol of Western culture's wealth and power.

Please don't misunderstand me. Al Qaeda members are certainly some of the nastiest people who have ever lived— they rank right up there with Jared Diamond's Maoris, who purposefully killed and ate and enslaved the entire population of Morioris. Certainly Al Qaeda must be exterminated, but by international police actions for their murderous behavior rather than by wars waged against countries that conveniently have resources we covet.

Currently war and defense are considered major business opportunities for many segments of the U.S. economy and for whole nations. Consider how much of the U.S. Federal Budget is dedicated to war, war preparations, and the aftermath of war. According to the War Resister's League, a group that traces its origins back to 1923, the U.S. spends 54% of its budget on war—18% for past military, 36% for current military, and an estimated $200 billion for Iraq and Afghanistan war spending. Forty-eight percent of global military spending is done by the U.S.! According to *Time* magazine[83], in 2008 the US spent $607 billion on to upgrade its armed forces, more than seven times the amount spent by our nearest competitor, China, which spent $84.9 billion. We are not a benign "big tree in the forest"—we are the most dangerous aggressors on the planet pushing the entire system to a self-defeating arms race. And most of the motivation seems to be creating business opportunities for transnational corporations at the expense of jobs for Americans.

There seems to be no limit to the profits companies like Haliburton and Blackwater (now XE) can earn in war. In our current insane system of constant warfare, the most adaptive nation is one that can evolve to make record increases in GDP in both war and peace time. Unfortunately there is a real-life example. By 2004, Israel had developed one of the fastest growing economies of any Western nation—they have positioned themselves as a shopping mall of homeland security technologies

---

83 "Breaking the Bank for War", Time, June 22, 2009, p 15.

as well as a fortress willing to defend from and/or attack any and all comers. According to Naomi Klein [84], they have lost their motivation for peace—their businessmen make the most money when there is unending, low-level, grinding conflict. Is this what we want as this world is getting smaller each year? This sort of "economic growth" can only be celebrated if it is measured by GDP rather than GPI as described in Chapter 3 and if sane moral standards are ignored.

War Is an Abuse of Ethical/Moral Behavior

As a social species with a moral foundation in our religious traditions, we're not supposed to murder, lie, or steal and we're supposed to take care of other people when they're having trouble. War seems to be a "free zone" for abuses of our global neighbors and nowadays also for abuses of our own citizens. And the aggressive taking of the resources from other countries is stealing, regardless of whether the resources are paid for or not because the citizens of countries we have exploited are left poor and/or homeless. How did all these social rules get suspended during war? Dropping bombs, shooting real or suspected combatants, and "collateral damage" to the lives, bodies, and lands of people and other species in conflict zones is not in any essential way different than acts of violence during robberies. The only difference is that one is sanctioned by the way our culture has evolved and the other is not.

It is our ethical responsibility as citizens of a democracy to control what is sanctioned by our government. The government, after all, represents us. To be ethical we must purposefully move

---

84 Naomi Klein, "The Shock Doctrine: The Rise of Disaster Capitalism," 2007, Metropolitan Books.

our foreign policy and military might toward peaceful (or at least limited) ways of settling disputes. There's a lot to be learned from the baobab story—those trees are massive, strong, and enduring and they are in no danger of losing their access to light. And if we're strong enough, the lesson of the Phoebes should be remembered also—we can endure some injustices in order to sustain peace. Our country can learn from models like that or we can dissipate our wealth, our land, and our children in war and war preparations. The lesson of the peaceful Moriori should not be forgotten. Leaving oneself defenseless and disorganized is a dangerous route to take in the world today. On the other hand, who can better lead the way toward peaceful conflict resolution than the U.S.?

Shifting from War-like to Peaceful

But to change our global impact from war-like to peaceful will take a fundamental shift in our culture. Taken together, the ideas of two scholars can provide a framework for moving forward that addresses all three of our time, species, global concerns for people, the environment, and the economy. The key to the fundamental shift we need is whether we use our wealth and power to promote the profit-driven dominating interests of huge, politically-connected, transnational corporations like big oil and agribusiness or whether we respect the diversity of cultures, ecosystems, and resources of each place on earth and thereby promote collaborations in trade between unique places.

The first scholar, Berkeley professor of cognitive science and linguistics, George Lakoff, has demonstrated that Americans and people around the world use one of two subconscious cognitive frameworks in various aspects of their lives[85]. Both frameworks are based on the family as the fundamental social unit that defines our core behaviors in contexts ranging from home to work to our social and political lives. One type of family is the strict father model—this is dominated by the head of the household who makes the rules and doles out rewards and punishments to household members as deemed appropriate. Family members are rewarded for obedience and for personal discipline that manifests itself as wealth/success. In two parent families, usually females rank lower than males and must follow the lead of the strict father. The second type of family is the gender-neutral nurturant parent model where Mother and Father together have core values of respectful, loving, supportive relationships with their children. In this model nurturing means empathy and responsibility—the parents must be strong to protect their children from dangers and to nurture good behavior in them. In shorthand, the two models might be called domination and collaboration. These frames have implications far beyond the family.

The second scholar, Vandana Shiva, is a physicist, philosopher, and feminist from India. India has strong patriarchal social and political systems that oppress women. India has also until recently been dominated by the exploitation common to Third World peoples by Western industrial nations. Because

---

[85] George Lakoff, "Don't think of an elephant!," 2004, Chelsea Green Publishing Company

of this background Shiva's classic writing[86] is strongly cast in male-female dichotomies but she has provided a philosophical framework in ecofeminism that can be easily adapted to our challenge to become more like the baobabs. In India animal husbandry, raising of food, seed saving, and recycling of crop waste for fuel and soil fertility has traditionally been done by women. The seeds saved for the next year's crop are from locally adapted plants in each small community that taken together represent a tremendous amount of genetic diversity within the crop species used in India. Food is raised in a multi-layered, high-intensity cropping system including ten or more food species in a single small field. The work hours involved and the products produced are not counted as part of the GDP of India because although this work allows people to make a living and feed their families along with preserving biodiversity, it is not paid labor and the products are not sold on international markets so the work, productivity, and biodiversity impacts are "don't count." Notice this is a collaboration model, collaboration between neighbors in a region as well as collaboration with the diversity and evolving power of natural systems.

According to Shiva, development in India consists of replacing the female collaborative model of people's relationship with the land with a male model of domination of the land, essentially the industrial mindset of uniformity and mass production that flies in the face of nature's operational rules from Chapter 5. The biodiverse small fields are replaced with monocrops of coconut palms (or other crops) for production of products that can be exported and grown with the aid of scientifically developed

---

[86] Vandana Shiva, "Women's Indigenous Knowledge and Biodiversity Conservation" in Maria Mies and Vandana Shiva, "Ecofeminism," 1993, St. Martin's Press

fertilizers and farm equipment. Biodiversity is lost along with the livelihoods of the people because by taking the diversity out of the system, mass production techniques can be used. Fewer people are employed, seeds developed in laboratories in other countries are planted uniformly over the landscape, and automated culturing methods are utilized. It is easy to measure the contribution to India's GDP of this agricultural system because it is simple to measure input costs and profit earned. But uniformity and mass production and simplicity are not the way of this ancient place. Domination of people and of the landscape does bring big profits to a few, but it robs nature of its diversity and people of their livelihoods as well of their locally adapted culture. For example, according to Raj Patel[87], 20 years after Green Agriculture was introduced to India, malnutrition had increased and the frequency of episodes of starvation had increased from once about every 100 years to once every 3-4 years. Now, of children under three years old, 46% are malnourished. Their reliable food supply has been lost to production of food for export and their GDP continues to rise.

Bringing Home the Bacon

In this slice of time in which we live, Western industrialized nations have habitually scoured the world for resources and brought them home to enhance the wealth of a few by providing material goods others will buy. Damage to other cultures and their environment has not been our concern because our personal I, me, now "bubble of unreality" has been secure until recently. Our country's military might and economic power

---

87 Patel, Raj. 2007. "Stuffed and Starved: The Hidden Battle for the World Food System". Melville House, Brooklyn, NY.

have been used to manipulate foreign governments into nearly giving away their natural resources in schemes that leave local people without the means to sustain themselves. U.S. foreign policy and military bases are now set up to secure "American interests" around the globe, much of which is an uninterrupted supply of oil and food to feed our ever-growing needs. This powerful machine of domination was not the original intent of our founding fathers. Their original intent was set up to stop that sort of exploitation of America by European nations and to ensure people living in the U.S. protection and freedom. Our culture and government have evolved from the Revolutionary War into a behemoth today that is spreading a culture willing to suck the world dry in return for an ever-growing GDP. The U.S. is now starting to get a taste of its own medicine.

This domination world view sees the U.S. as the measure of all value with space only for hierarchy and power instead of diversity and collaboration. This attitude translates into policies that give aid to other countries based on criteria that leverage their economies open to investment and exploitation by U.S. transnational businesses. In this worldview, the diversity of the world is seen only as raw material for insertion into a corporate monoculture of goods, agricultural practices, and interchangeable parts. With this kind of system, only a few of the most "worthy" (or maybe the most ruthless and amoral) individuals will succeed, and their choices will determine the choices available to the masses of people. Most people will descend into poverty and become the objects of continual disasters and alms-giving by the wealthy. Globalization in this context promotes massive money-making opportunities for a few by free movement of capital, technologies, and just about everything except laborers, who remain captive to the prevailing wages, environmental degradation, economy, and working conditions of their country. It is only recently with the onslaught of free trade agreements that Americans have gotten a taste of being on the receiving end of this domination version of globalization.

From the standpoint of the non-human species around us, the current concept of globalization is a disaster. Species that have previously been constrained in their evolved communities by disease and predators are now being moved haphazardly around the globe with agricultural produce, packing materials, and the pet trade. In their new homes, ecological constraints are released and pests like Zebra mussels, White Top, and Emerald Ash Borers can reproduce to their fullest biological potential. As the populations of invasive species boom, native species are decimated. For example, currently Emerald Ash Borers are killing Ash trees in the Midwest, from Virginia to Wisconsin to Missouri. In 2002 they accidentally entered the U.S. in the wood of packing pallets and are spread as people move firewood from dead ash trees. Similar stories can be told around the world. Hawaii's unique native birds are almost extinct because of predation by Brown Snakes, which can even travel in the wheel wells of airplanes. Free movement of unsterilized goods and conveyances is rapidly shifting whole ecologies toward global uniformity. Invasive species hop scotch around the globe because we have not respected the importance of diversity in the on-going evolving story of the earth. Things don't have to be this way. Globalization is a good thing if we go about it in an ethical way. Trade partners rarely make war on one another—both partners have too much to lose.

Ethical Globalization

The ethical way to go about globalization is by using a collaborative world view, an experience humanity has not enjoyed.

In a collaborative model of globalization, the diversity of cultures would be seen as valuable differences to be promoted and cherished and nourished as potential trade partners who have developed innovative ideas applicable to other situations. The cooperative worldview would nurture and protect the microorganisms, flora, and fauna of the different ecosystems so the species could continue their local evolution. Unsterilized food or materials would not be permitted to move between nations because this is the primary route for ecosystem-destroying invasive species to move. Different foods and food systems, each locally adapted to the culture and environment of the place, would be promoted and protected, especially those that continuously improve local cultivars of crop plants and animals. In this system, the hand labor associated with local food production would keep much of the human population employed, especially in third world countries. Information, art work, music, technological innovations, and cooperation could spread and be adapted to the local mores and needs without strictures that force uniformity. Global progress could only be measured in this context by sophisticated metrics like the GPI rather than GNP. Ideally globalization that is ethical from a time, species, global perspective would consist of community-to-community and business-to-business partnerships where both partners mutually benefit from the trade.

Notice how different the two models of globalization are—the domination model that demands military intervention to secure American interests as we exploit the resources of other countries for our benefit versus a collaborative model that nurtures countless collaborations around the world where both partners benefit. The potential of war with the latter model becomes minute and selective pressures to develop murderous groups like Al Qaeda are mitigated. Local ecosystems are protected and each innovation has the potential to be spread around the globe and then be adapted to other local conditions and needs.

Important Concepts of this Chapter

All species have conflicts but war, collective attacks on members of one's own species when females and offspring are present, is restricted to chimpanzees and humans among mammals. This uniqueness was caused by the feeding ecology of the chimp-human ancestor in Africa that resulted in isolation of females and concomitant male aggression.

Among humans, warlike societies develop because of their cultural evolution's response to their feeding ecology. These warlike societies easily overwhelm and kill members of peaceful cultures.

We assume war is a part of life like birth and death and war is considered a "free zone" for breaking many moral rules. But war and war preparations have created massive environmental damage on this small planet, so an ethical framework must seek alternatives that don't make our civilization vulnerable to warlike factions or countries.

There are models in nature that allow strong species to maintain their strength and safety without stimulating arms races in the competing species around them.

A collaborative rather than exploitive, domineering style of globalization would take the world toward a more peaceful future. Trading partners, with equal benefits and rights, do not make war on one another whereas exploited, threatened groups become violent easily.

Globalization should be designed so biological organisms are not moved because they become invasive and cause the collapse of host biological communities. Globalization should also be designed to avoid monocrops of single technologies and/or crops because this kind of simplicity is unstable in the long term. Innovations should spread easily around the globe and act as mutations in their new community—application of the innovation would then become locally adapted.

With millions of community-to-community and small business-to-small business collaborations as the core feature of our globalization, few people would be willing to make war on anyone else.

* * *

In chapter 1 of this book, I have attempted to communicate the importance of understanding earth from a time, species, global perspective so we can make ethical decisions that will allow humans and our non-human neighbors to persist on this planet for the long term. Chapter 2 was a short trip into the nested levels of influence in American culture that have turned us into obligate mega-consumers. Chapters 3, 4, and 5 discussed how breaking out of the assumptions of our culture can allow us to step mentally back into being a part of nature, subject to its rules, but also with permission to modify our environment for our comfort and well being as long as the technologies we accept respect all of nature as key to our well being. In Section 2, I have discussed four key elements of the systems around us that impact us: the concept of the commons with its susceptibility to damage using the oceans as an example; soil and water and our food production practices as key commons that underpin our lives and impact earth's life support systems; land use practices as the foundational policy that controls the structure of our communities and the health of the non-human biological communities living on the land with us; and finally, a view of war as an outcome of exploitive cultural behavior rather than mutually beneficial cooperative behavior.

With this background, I now invite you to the last part of this book, which deals with how life can be if we approach it with an ethical time, species, global perspective. If we purposefully live our lives and run our businesses and use our votes thoughtfully and think green within the rules of nature, then our choices can lead to an abundant and healthy future for ourselves and for the rest of life on this planet. The physical and spiritual rewards of this approach to our personal responsibility as the most intelligent species on earth are worth every bit of change we must accept to reach this worthy time, species, global goal—in the process we can meet our own I, me, now goals more effectively.

Section IV

Freedom and Prosperity

Chapter 10

Making Choices in Our Personal Lives

One of the themes of a time, species, global perspective and green thinking is diversity and local solutions. Those ideas are of paramount importance for this chapter because there are countless paths toward individual freedom and prosperity, all of which work within nature's operational principles. The fact we are all different and have the freedom to pursue our lives, liberty, and happiness in this country is a key aspect that makes the U.S. the best place in the world to live. Just as the abundance and diversity of nature is based on fundamental operational rules, there are fundamental principles we can use as we make our individual decisions as we use our time, species, global perspective to live within nature's operational rules.

Of course, we need a way to track how well we're following these fundamental principles that help us live as a part of nature while maintaining and improving our quality of life. Here I'm talking about a quality of life that would make a person on their death bed feel like they've done a good job of making their choices: they've lived fully and enjoyed their life; they've raised children who are well adjusted and can cope with the changes the world throws at them; they've maintained their health in so far as possible by taking care of their body and spirit; they've improved their local and global community; and they've left a swath of health and abundance behind for all living creatures as they've gone about the business of making a living and living a life. So how can you track such lofty goals?

My suggestion is to watch your trash, the discards from your life. The trash I'm talking about isn't just what you discard at home and your workplace each week. I'm also talking about the full life-cycle unused and unusable waste associated with everything you are buying to meet your needs and wants. For example, an empty spray can of wasp killer in your trash will help fill a land fill but also involves the production of the can, spray nozzle, cap, transportation of the product to you and after you're done, as well as the chemical feedstock for the poison inside. Each spot where production has occurred means resource depletion and pollution of various sorts. If you live in a city, instead of the wasp killer, you can use Aquanet hair spray; it glues the larvae into the wasp nest so the young can't hatch. Aquanet isn't a good thing to use on your hair though because it's explosive when you touch a match to it. If you live in the country, just plant flowers that attract bees. Bees are natural enemies of wasps—the wasps are soon gone if you have enough bees plus you get the honey and wax they make.

That's just one example. Thinking green means you have to do some thinking and learning before taking action. It's amazing how much money you can save and how you can find win-win solutions if you investigate before buying. The idealistic goal I advocate is to reduce your unusable and pollution-generating waste to zero and to ensure all your other waste joins a "waste as food" stream. When you look at it that way, much of your waste can be used to reduce the things you need to buy. For example, composting reduces your need to buy fertilizers.

It's amazing how much of our wealth we throw away with the trash. One day I was listening to PBS and they were talking about the ten richest women in the world. Nine of them had become rich by marrying old, rich men. The tenth was a Chinese woman who found value in U.S. trash. She recycles paper, makes it into packing materials, and sells our waste back to us. As I see the giant rolling trash containers out for garbage pickup each week in my neighborhood, I wonder how much more of our wealth we are throwing out with the trash. I've got mine down to one trash bag every 3-5 months. That bag contains mostly plastic bags that have been reused several times before I discard them. You may wonder what I do with the rest of my trash. The simple answer is that I don't buy it to begin with. If I can't avoid buying it, I choose containers that can be recycled or reused. Finally, I compost all organic waste using compost worms and a Compost Tumbler©. Don't laugh. It works and it saves me lots of money!

The Secret of Abundance

There is a wonderful essay by Eknath Easwaran[88] called *The Lesson of the Hummingbird* in which he compares how hummingbirds make a living with how Americans have been making theirs. Hummingbirds are attracted to beautiful flowers, fly in front of them and sip nectar using their long beaks to sip from the bottom of the flower, pollinating the flower in the process. When the hummingbird leaves, both the hummingbird and the flower are better off as a result of the interaction—the hummingbird has gained the food it needs for survival and reproduction and the flower can now produce seeds for the next

---

[88] Eknath Easwaran, 1990, "The Lesson of the Hummingbird", from "In Context", included as a reading in "Discussion Course on Voluntary Simplicity", 2003, Northwest Earth Institute, 506 SW Sixth, Suite 1100, Portland, Oregon, 97204

generation of its species. Notice the way the hummingbird feeds *increases* the abundance of its food. It can eat as much as it wants and the benefits just keep growing. And this goes on all over the world. Many of the most beautiful flowers on earth owe their attractiveness to the selective pressure placed on them by the various species of hummingbirds that pollinate them. In turn, hummingbird beaks come in a wide variety of shapes and sizes in response to the evolution of the nectar-bearing flowers. Diversity and local adaptations in the flower-hummingbird interaction have enriched the beauty, biodiversity, and abundance of thousands of places around the world.

In contrast, with our present ethic, if a human wanted the nectar in the bottom of flowers, we would cut down the flowering plant, rip off the flowers, extract the nectar using toxic compounds, stomp on any living shoots, and leave the dead plants as a pile of toxic waste. Humans would get only a little bit of nectar from each flower so it would be necessary to kill whole fields of plants around the world to meet even a portion of our need. Our way of collecting things like nectar damages the ability of the rest of nature to meet our needs and increases resource depletion that will ensure a future of deprivation. We've been taking more than we need, wasting a lot of what we take, and destroying nature's ability to evolve to meet our needs.

The lesson of this comparison is clear—if our time, species, global ethical imperative is to adopt behaviors that ensure an abundant future for all living things on earth, then we must learn to take what we need in clever ways so we create health and abundance for ourselves and the species around us. Obviously this necessitates a sea-change shift in industry (covered in the next chapter) but from a personal standpoint, the first step toward this goal is learning to be purposeful individuals who thoughtfully and carefully design our lives to meet our personal goals. This means freeing ourselves from living paycheck-to-paycheck and freeing ourselves from the spiritually expensive "cheap" practices of our culture. To be spiritually and financially free, we must make choices that free us from obligate consumption (Chapter 2).

What Matters Most to You?

The Center for a New American Dream[89], The Simple Living Network[90], and a movement called Voluntary Simplicity[91] are three related organizations that have developed sophisticated paths forward for recovered obligate consumers. The idea these groups espouse is similar to Dave Ramsey's budgeting to overcome debt in order to regain financial freedom. The theme of voluntary simplicity is to thoughtfully de-clutter your life of everything except those things that matter most to you.

De-cluttering your life can start with material goods—if you give up your $30,000 SUV, you also give up its payments and gain control of the time you spent working to make the car payments, the higher insurance, and to buy the extra gas and oil for such a large vehicle. If you move to a home close to where you work, you give up your former "American Dream Home" but you also gain personal time you formerly used to commute to and from work. If you give up the convenience of packaged frozen dinners and other prepared foods, you gain exercise by cooking from scratch, save money because basic ingredients are much less expensive than prepared food, and you have the

---

89 http://www.newdream.org/

90 http://www.simpleliving.net/main/

91 http://www.greatriv.org/vs.htm

luxury of eating chicken that tastes like chicken and knowing what you're eating. If you remove all of the cheap junk from your home, you can live in a cleaner, more restful space that takes less time to clean and has fewer allergens present. If you are super-selective in what you buy to furnish and decorate your home, you can live with timeless durable objects of beauty rather than piles of clutter than offend any sense of inner harmony.

I hope you see my point—curing obligate consumerism doesn't mean austerity or deprivation. Instead, it means freeing ourselves to actively choose a purposeful path for our life.

Trade Offs

1. If you have been eating a corn-based industrial food diet, you may have forgotten what grass-based (aka real) chicken, pork, eggs, cheese, milk, and steak or wild shrimp tastes like. Seeking out and spending the extra money for grass-based animal products produced as close to where you live as possible, at least part of the time, will give you a sense of what food should taste like. Amazingly, chicken actually tastes like chicken! Your health will benefit from every bite of real food you eat because the balance of fats in grass-based animal products is healthy for humans whereas the fats of industrial food are a major factor in the development of many common diseases in the US. Grass-based food is superficially more expensive than the federally subsidized industrial food but by applying the "waste as food" principle in your own kitchen, you can substantially reduce the cost. For example, rather than just throwing away that steak bone or chicken bones, you can simmer them for a few hours with some herbs, pick the meat from the bones and cartilage,

add some carrots and other vegetables, and make a delicious soup for lunch.

2. If you buy produce from the supermarket, you may have forgotten what a real tomato or strawberry tastes like. But if you start growing your own vegetables and preserving them for winter, you can regain some of that flavor and food health in your diet without much expense. In the country, a garden is easy—just give up a patch of lawn and fence it to keep the woodchucks (!!!), deer, and rabbits away. In the city, community gardens are an option. For example, in Seattle the Magnuson Community Garden at the Warren G. Magnuson Park has 135 P-Patch plots for individual gardening in the context of a park that also has an amphitheater, a children's garden, a native plant demonstration area, a native plant nursery, an orchard, and a tranquil garden. This project has been so successful that other communities in Seattle and around the country are developing similar parks. With a garden, you will also have incorporated exercise into your life and have regained a connection with the earth for yourself and your children as your source of food.

3. In big cities, getting locally produced food is difficult but this may not be the case in the future. According to *Science News*[92], Dickson Despommier and other researchers, architects, and city planners are looking to incorporate vertical farms into cities—tilapia on the first floor of a high rise energy efficient greenhouse and tomatoes on the 12th. A vertical farm of this sort would use the waste of the city (CO_2, treated sewer water, compost) and the waste of the

---

92 Ehrenberg, Rachel. 2008. "Let's get Vertical: City buildings offer opportunities for farms to grow up instead of out". Science News, October 11, 2008.

fish water to grow food plants using natural fertilizers. This sort of farm would make use of companion planting to enhance the health of crops and their resistance to disease. One illustration in this exciting article included peas, spinach, thyme, tomatoes, Brussels sprouts, peppers, chives, lettuce, peaches, apples, cabbage, strawberries, and cherries all on one floor! I'm sure they could grow the Chilean grapes so many people are fond of too!

4. If you release your children from an endless round of soccer, band, martial arts, etc., and encourage them to choose one favorite activity, then you will also free them to live as a child instead of as a corporate-executive-in-training ruled by an external schedule. Think about it. You could spend quality time with your child in your garden helping them to connect with the source of their food. In humans' long history, children learned all of the skills they needed to be successful members of their community from the adults around them instead of in formal learning situations like our modern industrial-model schools and like formal children's activities. Reading a book you and your child both enjoy can do more for their reading ability than hours spent in school. And children need time to just play and explore the world around them. One of my most precious childhood memories is the time the neighborhood children and my siblings and I spent building a tree house in a big, old Oak tree. We had a president, vice-president, treasurer (for ~27 cents of treasury), and a secretary. Being a girl, I was the secretary. We scavenged old lumber and shingles from the farmer across the road and built a sturdy tree house about 20' up in the tree. Nowadays, many places cut the lower limbs off of trees to keep kids from climbing to protect them

from falling. Instead why don't we just teach our children how to climb safely?

5. We're all short of time but if you turn off "video"—including TV, Netflix, You Tube and the like—for most of each day, you free up your time for reading, family interaction, and other activities of your choice. We've become a civilization addicted to passive forms of entertainment, an addiction that can be measured by our expanding waistlines. Alternative activities that incorporate exercise are easy in the country but can be more challenging in the city. Still, to quote an old adage, "Where there's a will there's a way." Probably it is most difficult for those who live in a small apartment in a major city. Technology could be a help there. Have you seen the Wii software and equipment? It's possible to bowl, balance a boat on a river trip, play tennis, do yoga, hula hoop, etc using Wii Fit or Wii Sports or a multitude of other active and interactive forms of recreation. The bottom line is that healthy, happy people are that way from having the time to be involved in active leisure time activities.

6. Voluntary simplicity doesn't stop there. As you de-clutter your life, you can also regain control of where and when you work. Ideally you would work for a truly green employer—there's more information on them in the next chapter. Many Americans are tied to jobs they hate so they can meet their monthly payments for cheap junk. A while ago I heard about a family—Mom, Dad, and one child—who had three microwave ovens so they could prepare their Swanson dinners simultaneously. This saved ten minutes they could use to eat together. I think that story is more tragic than situations I've seen in Madagascar where there was just enough food available to feed the whole family their meal of rice.

At least the Malagasy family had time to sit together while the rice cooked and then to eat it in peace. Rushed meals of indeterminate factory-farmed food laced with sugar and fat cannot nourish the body properly let alone the spirit or the family. When you think about it like that, I suspect it is easier to consider giving up a high-paying, impressive job that comes with an expected lifestyle of Gucci loafers, Armani apparel, Louis Vuitton luggage, and a Rolex watch in return for a life that allows you time to live. Or, if those material goods are important to you, at least choose an employer who is thinking green (not just green washing) so you can feel a sense of pride and ownership in a bright future.

The Fly in the Ointment

There is a fly in the ointment of Voluntary Simplicity and our ability to fit our work lives to the life we wish to live. That fly is the health care industry in the U.S. Most Americans have health care insurance tied to their work—the bigger and flashier the job, the better the insurance, usually. If Michael Moore's documentary, "Sicko," is correct, then at least subconsciously we Americans are deeply fearful of losing our jobs because we lose our health care coverage with the job. Many jobs now, even good jobs, do not come with health insurance and the costs associated with even relatively minor illness can quickly skyrocket out of control, even for those with insurance. Currently, 17% of the U.S. GDP is given over to health care costs, much more than any other developed nation. Current trends put our health care costs on track to reach 20% of GDP by 2016.[93] Health care

---

[93] Poisal, J.A., et al, "Health Spending Projections Through 2016: Modest Changes Obscure Part D's Impact". Health Affairs (21 February 2007): W242-253

costs are rising at more than two-times inflation, a situation inherently unsustainable. At some point in the near future, the system must implode because people have needs beyond paying for health insurance. Frankly, I haven't found any way to simplify my way out of the health care problem but the Obama Administration is starting to give me the hope that uniquely American solutions are available.

Still, it seems to me this is a primary area where we Americans need to rise up together to put a strong selective pressure on this evolving part of our society. Our medical insurance Godzilla needs to be shrunk back down to its proper place in our culture—to put things in perspective, public health projects like sources of clean water and programs of immunization have saved more lives than our health care system ever has.

Along with excessive corporate profits, many of us have a part in the excessive cost of our health care system because overconsumption of health care services is partly responsible for making our system the most expensive and wasteful in the world. This overconsumption is not in our best interest because what have become standard U.S. medical care practices are not only expensive, they also imperil the health of patients. Nearly 100,000 people per year die of medical errors, sometimes errors made during procedures that are unnecessary[94]. Do we really need all those MRIs, full-body CAT scans, colonoscopies, diagnostic X-rays, and heart catheterizations? Shannon Brownlee says not[95]. As a matter of fact, these invasive tests have false positives and false negatives. When false results are coupled with

---

94 Shannon Brownlee, "Overtreated: Why too much medicine is making us sicker and poorer," 2007, Holtzbrinck Publishers.

95 ibid

the occasional errors our doctors make in diagnosis, there is an *increase* in diagnostic errors. We would have been better off just trusting the judgment of our doctor. This really hit home with me when the husband of a friend of mine spent more than a year thinking he had prostate cancer due to a false positive PSA test. That was a tragedy, both from the standpoint of the emotional health of the family but also from the standpoint of the dollars spent treating his non-existent cancer.

William McDonough's question "What are your design goals?" should be central to any discussions about modifying our health care system and our relationship to it. Our health care system should be designed to improve and protect the health of our citizens at the lowest cost possible. And we citizens should use the system sparingly, just enough to ensure our individual health. Of course, such a system would be a commons of sorts so we would need definite social arrangements to keep the system from being abused through ignorance or greed. The system should be simple and easy to use for even the most uneducated among us and it should be accessible and affordable for all of us. Any element of the system that doesn't contribute to those criteria should be selected out of the mainstream health care system—that is, it should be sent "extinct" as being non-adaptive.

You may respond that you have no power over this major sector of our society. But you must remember where this sector gets its money and power—from us, the citizens. If we quit submitting to unnecessary expensive tests and start trusting our doctors, then the health care system loses some of its power and money—along with that influence over public policy is reduced. This would also be a good time to make sure your elected officials at all levels of government know you are watching their efforts to repair the U.S. health care system. Together, these actions would help develop a more positive trend for the future even if a full solution may not be immediately possible. Taking better care of the environment would help a lot too.

Simplifying Environmental Concerns

A key piece of thinking green with a time, species, global perspective is to simplify how you take care of environmental concerns. We are constantly barraged by all of the things we should be worried about with respect to the environment and all of the endless things we should do to save the environment from destruction. It's easy to dissipate our time and energy on efforts that don't make any significant difference in our (human) ability to live on Earth for the long term.

A case in point is the campaign that has occurred in our schools for picking up litter and recycling. Litter is unsightly but it isn't going to cause Earth's life support systems to crumble. Recycling is fine and an important thing to do, but one wasteful neighbor can use up in one week all the resources you save in a year. This is a classic "slobs" and "suckers" commons dilemma— the only way out of it is to work toward mandatory recycling in your community or, better yet, put the pressure on corporations and elected officials to limit packaging and to produce durable, truly green products. In other words, your time is better spent changing the system in which your community functions rather than sacrificing your time and energy by picking up litter. In the meantime, I hope you keep recycling and picking up litter, but don't kid yourself about saving the earth by doing it.

Too many little things can overwhelm people. For example, I have a nice little book called *50 Simple Things YOU can do to Save the Earth*[96]. I don't know about you, but I have way too much to do to keep 50 things at the front of my brain. Instead, I recommend building your consumption patterns around the results of a scientific study done by Michael Brower and Warren Leon of the Union of Concerned Scientists[97]. Their results clearly point to our ability to make significant improvements in the environment by being mindful when we make major purchases that accrue environmental benefits as we go about living our lives. The logic is that one big, right decision is worth more than thousands of little right decisions and it is much easier to accomplish. Here's the logic of their recommendations.

Determining the Big Things

Brower and Leon[98] started out by determining the most threatening environmental problems—those with the greatest danger to human health and earth's ecology. To make these critical choices, they used two comprehensive studies on risk assessment, one by the EPA and the other by the California Comparative Risk Project. These studies had collected all of the available scientific data on the health and ecological impacts of a wide range of human activities and ranked them by impact. Brower and Leon selected only those problems that were ranked as medium or high risk in both studies and then they dropped any problem that had already been addressed (eg. stratospheric

---

96 The Earth Works Group, "50 SIMPLE THINGS YOU CAN DO TO SAVE THE EARTH", 1989, Earthworks Press, Berkeley, CA

97 http://www.ucsusa.org/

98 Michael Brower and Warren Leon, "The Consumer's Guide to Effective Environmental Choices", 1999, Three Rivers Press.

ozone depletion). This netted them four problems: air pollution, global warming, habitat alteration, and water pollution.

Next Brower and Leon built a model to link consumer purchases and activities to specific environmental problems. In this model, they recognized that each purchase and/or use of a product could have both direct and indirect impacts on the environment. For example, when you spread a fertilizer on your lawn, it has a direct impact on nearby water pollution, and it also has an indirect impact on air and water from manufacturing the chemicals. As a second example, eating beef raised in a Nebraska feedlot has a huge impact on water resources from contamination by animal feces as well as on pesticide pollution from the corn fields used to raise corn to feed the cattle. In addition there is also pollution from the manufacture of materials used to raise the corn. It's amazing how much pollution is packed into one industrially-raised steak when you think about it like that!

The indirect sources of environmental damage were then distributed among about 500 industry and commodity sectors they studied. Brower and Leon combined the amount of damage with data on dollars spent by the consumer to determine environmental damage per dollar of economic output. To translate these data into consumer purchases, Brower and Leon used an input/output model commonly used in economic studies in each industry sector for any given set of consumer purchases. Lastly, they combined the direct impacts of consumer use of products with the indirect impacts associated with their production to get total environmental impact for various activities. This whole thing translates into a fairly simple set of data—it quantifies environmental damage per dollars spent.

This sounds complicated and indeed it was, but the messages coming out of their detailed analysis are fairly easy to digest and use. The standard they used to develop their recommendations was impacts associated with specific activities that were more than five times the average impact of consumer purchases. For example, some obvious high impact activities include power-boats, pesticides and fertilizers, gasoline-powered yard equipment, recreational off-road driving, and the use of hazardous cleaners and paints. Making the choice to focus on activities with especially high impacts allowed Brower and Leon to get to the key messages from their data. Following are their core messages expanded by examples.

As you read the following messages, you may feel like this information is all well and good, but you have already built a life and there isn't room to make these kinds of choices anymore. You may have a two career household or may need to live near an aging relative or you inherited a paid-for SUV in a divorce and can't afford a car payment. I sympathize but I also encourage you to become aware of how these past choices constrain your life. I've heard it said that wisdom consists of changing the things we can control and living with the things we can't control. Remember the story about the cheese in Chapter 1? It's important to recognize when your cheese is getting really stale and to make the changes necessary so you can enjoy new cheese. With that in mind, let's look at the ideals. Possibly they will give you ideas that adapt to your situation to help you improve your life.

Big Ticket Item #1: Choosing Where to Live

The most fundamental choice you can make to reduce your environmental impact and to gain control of your life and time is choosing where you live. Living close to your work and your children's school changes your whole relationship with time, the roads, and ease in securing your basic necessities like food. To accomplish this vital choice, it's important to think long-term because, as your children develop, they change schools and your employment may change as you go through life. This means your situation will evolve through time and it's easy to get caught in a life that is sprawled all over requiring you and your children to tithe time and money to the almighty god of cars and the road. When you find yourself in that situation, I highly recommend re-calibrating the choice of where to live if you can afford it.

Tangential to where you live is the question of how you ensure your mobility. Rather than falling into the trap of purchasing a SUV (also known as a SAV—Suburban Assault Vehicle), a huge expense environmentally and in your personal budget, choose the most fuel-efficient vehicle you can find that meets your needs. Many of us think our children will be warped if they must ride in a small vehicle, but that really isn't the case. I think back to my childhood when there were eight of us crammed in a small, paid-for car. I'll never forget the day my mother stopped at the local grocery store to buy day-old bread. The grocer had a small smile on his face as he said he would sell it to her if she took it all. We ended up with our faces pressed against the windows in order to fit all the bread into the car. My siblings and I are all fairly normal people so I suspect no harm was done.

If you find you are living in a place that makes it so you have difficulty without purchasing a second car, think twice about moving closer to the things important to you. That would be a better alternative rather than putting your time and treasure into another vehicle. Even if you purchase a second vehicle, it's important to get in the habit of organizing your life so you make the fewest trips possible. Plan out your route in advance and avoid running here and there to get just one thing. If you have made a wise choice in where you live, you will also have the option of walking, bicycling, or taking public transportation for most of your travel. I've even seen a fellow who chose his home in a place that allowed him to kayak to work and back each day—turning his commute time into recreation and fitness time. In Chicago, research has shown that property values increase with proximity to public transportation—if you ride public transportation, you can use the time to start your work day rather than spend the time driving.

Above I've made some suggestions, many of which may not fit you. Nevertheless, please notice that the choice of where we live drives our environmental impact and also the structure of our lives. When you think about it that way, it's easier to spend considerable time studying the situation carefully before making your next home-buying (or renting) decision. Certainly, superficial motivations like being attracted to a particular house or landscaping or apartment layout should not be the primary force in making so fundamental a decision about our lives.

Big Ticket Item #2: Energy Efficiency

Almost as important as where you live is what you choose to live in and how you power it—that is, the choice of your home. Here I'm talking about sizing a home to your needs, making sure it is built to the highest energy standards, and built with materials that are not toxic to your family. Notice I am not mentioning whether the countertops are Corion or granite nor whether there is a hot tub on the deck—these aspects of a home are superficial and can always be changed. Many builders today sell houses and condos based on the feel you get as you walk through the place—a "new home" smell is considered attractive but you should know that any smell of that sort is an indication of toxins leaching out of the building materials. Many so-called "green" homes on the market are poorly constructed with methods and materials that will add to your monthly energy bills, repair bills that start shortly after you buy, and health impacts that increase through time.

Houses built to code are nowhere near the kind of quality you should look for in a home because the standards are based on political compromises rather than any true measure of what is best for people. My son Darren bought an expensive home built to code in a Denver suburb—it was beautiful with hard wood floors and sky lights and all sorts of nice amenities. It wasn't long after he purchased it that he realized the windows were of such a poor quality that the house's energy inefficiency was dramatically *increasing* his heating and cooling bills while simultaneously *decreasing* his family's comfort because it was chilly from drafts in the winter and hot in the summer. As long as he owned the place, he tithed monthly to the energy companies for the builder's lack of attention to crucial details. More recently, Darren bought a home in Sheridan, WY, that had the same problem. This time he replaced all the windows after living in the house for a winter. Because he's an engineer, he quantified his savings. Using normalized data, he found a 40% savings on his energy bill plus his home is now much warmer. Clearly, it's worth the time to seek out a home built to standards that will shelter and protect you at the lowest possible operating cost. Fortunately, it's not all that hard to do now.

One sure way to buy a home that is toxin-free and built at the highest standards is to buy (or build) a LEED[99] certified or

---

99 http://www.usgbc.org/DisplayPage.aspx?CategoryID=19

Green Built[100] home. Builders of these types of homes work from a check list of options that can be used to reach the highest levels of energy efficiency and to use locally produced, non-toxic building materials. The work of the builders is inspected during the building process and certified by the LEED professionals or the Green Built professionals. This process makes a huge difference in your home. Recently, Habitat for Humanity in the Grand Rapids, Michigan, area built a LEED certified home for a large family. This family is now cozy and warm in their new home and paying only an average of $36.00/month for its entire energy costs. Accomplishing this meant building the structure to standards far beyond building codes and installing energy- and water-efficient appliances.

Last year one of my former Environmental Ethics students stopped into my office to visit with me about the changes she had made in her life as a result of hearing the suggestions and information I'm now sharing with you. Rae had been living in a small apartment sited some distance away from where she worked as a barista for Starbucks and also far from campus. In the year after graduation, Rae purposefully rented a larger, LEED-certified apartment that had higher rent than her old apartment. The savings on her energy bill more than made up for the increased rent. At the new apartment, there was clear story space where she grew garden vegetables in pots plus the new apartment was located so she could walk to work, public transit, and to most of the places she wanted to go. The savings in obligatory expenses were allowing her to save money for her future goals. This young professional was happy and excited

---

100 http://www.greenbuiltmichigan.org/; This is the Michigan organization—other states have similar organizations, many under the name of Built Green.

about life and her future—what a change from the rushed, stressed student who had been in my class!

If you buy an older home, it's worthwhile to buy a smaller home than you can afford and then use your extra money to turn the home into the most energy-efficient place you can. If you already own a home, you can upgrade its energy efficiency and start saving money. A recent article in *Time* magazine called *Greening This Old House*[101] highlighted some important information about energy efficiency. In the U.S., 43% of our carbon emissions come from powering buildings and 50% more energy is used per square foot to heat pre-1939 homes versus post-2000 ones. And we've already seen through my son's experience how poorly most post-2000 homes work. The first thing you want to do to it is to insulate your home far past building codes, making sure ventilation to the attic is properly done. If you can't afford that, you can still get considerable savings by making windows, doors, and basement walls as air leak-proof as possible. A tube of calk is worth its weight in gold when applied carefully to plug up the air leaks. I've even stuffed old rags in drafts I've found around my basement windows. If you can afford them, new energy efficient double- or triple-glazed windows with low e (low emission) glass are an excellent investment. Your old windows could be used to set up cold frames in your garden rather than throwing them away in a landfill. The goal is to make your home leak-proof so it keeps your family comfortable and saves you money that you can use in ways you *want* to spend money.

When you use less energy you can choose your energy supply to be generated in ways more friendly to the earth. This

---

[101] Walsh, Bryan. 2009. "Greening This Old House: Saving money and the planet by upgrading older homes". Time, May 4, 2009.

choice puts a selective pressure on the U.S. energy market, moving it in a direction that is less harmful to the earth. If we all refused to buy nuclear and coal energy, the market for energy would change in short order. The energy source for my home is completely electric with a 100% green energy source. You can opt for green energy from most major utility companies but currently green energy is about 20% more expensive than coal and nuclear even though these traditional sources cause dramatic environmental problems. I've solved the ethical dilemma by offsetting the increased cost of green energy by minimizing my energy use. One key way I've done this is by installing a geothermal heat pump for heating and cooling. Other individual renewable energy options on your home include wind and solar. Installing equipment of this sort knocks the top off your energy bills permanently. You'd be amazed at how much money you can save with energy efficiency like this even paying the extra for green energy.

In 2008, my heat pump needed repair after more than ten years of trouble-free service—my home was on electric back-up heat for 2 ½ weeks in February, a fairly cold month in this part of the country. During the 2 ½ weeks, I lowered the temperature of the house, burned my Hearthstone wood stove constantly, and still had an electric bill nearly $100 higher than is normal for me. I tell you this to give you an idea of how much of your monthly income can be saved for yourself if you make the ethical decision to choose green energy and energy efficiency even if we now have to pay a premium for those choices. In the future, I expect polluting sources of energy will have to count their damage to our health and security on their balance sheets so green energy will become less expensive than traditional energy. Still, there are other benefits in addition to lower bills.

These kinds of choices also give you peace of mind. For years, I have felt no concern when the newspaper announced home heating oil had increased in price or natural gas prices were higher—it feels great to be free from the vagaries of the fossil fuel market.

Big Ticket Item #3: Energy Star Appliances

Consider replacing old appliances (including toilets and hot water heaters) with state-of-the-energy-art. You can tell this kind of appliance because they will be labeled Energy Star. One thing we do poorly in the U.S., Energy Star or not, is supplying hot water. The system now used in the U.S. is one of the most energy-inefficient, wasteful inventions of our time.

Right now, no matter where you are, there are 20-60 gallons of water being kept hot just in case you might want to turn on a faucet to get warm water. This is occurring when you are at work or sleeping and even when you are on vacation. Heating water is thermodynamically inefficient so this wasteful practice is a major consumer of energy. The system doesn't work all that well either—if you have company, the last people in the shower end up taking a cold shower. Also, each time you turn on a hot water faucet, you must first run the water out of the pipe before you get any hot water plus you must leave unused hot water in the pipes to cool off. Waste at every turn and environmental harm at every turn also—wasted water, wasted energy, and wasted time waiting for hot water. Why not install a small water heater in every room that uses hot water, but produces hot water only on demand and never runs out? Actually, this kind of water heater is already being used in Europe and is available but not widely used in the U.S. One brand I checked recently costs between $350 and $700 depending on how big a tankless hot water heater you need. That's more than a regular hot water heater, but think of all the savings you will make in your energy bills over the years. I just wonder why these sorts of things aren't regularly installed in every new house built and in every old house undergoing renovation.

Big Ticket Item #4: Cleaning Products

In addition to energy, you can protect your health, save money, and enhance the lives of your non-human neighbors by choosing nature's way to keep your home and yard clean and tidy. First, find non-toxic, inexpensive cleaners to use—some combinations of commercial furniture polishes and cleaners create fumes toxic enough to kill pet birds, but most household cleaning can be accomplished with baking soda, vinegar, and basic soaps. To keep your home free of stray toxins and to keep your local water clean, avoid the use of fertilizers and pesticides in your lawn. Become locally adapted by landscaping for the climate of your region—if you live in a dry area, xeriscape with drought-tolerant native plants, and if you live in a wetter area that will support grass, keep the grass to a minimum so it can be mowed with a push mower. Gas powered lawn tools create ground-level ozone that is damaging to your heart and lungs—they also disturb the peace of the neighborhood, so they shouldn't be used if you can avoid it. Gerald Celente[102], head of a think-tank of gurus who monitor trends around the world, predicted in 1997 that edible landscapes would become more common in the 21st century. With the rising cost of fruits and vegetables, why not use most of your lawn to grow food? You

---

102 Gerald Celente, "Trends 2000: how to prepare for and profit from the changes of the 21st century," 1997, Trends Research Institute

can eat it or preserve it fresh from the vine and you can even become an agent of natural selection by saving and using the seeds of the varieties that grow best in your local place.

Big Ticket Item #5: Moral Meat

Now let's do a little thinking about what you can do inside of your well-sited, energy-efficient, non-toxic home. One thing most dear to my heart is eating, an activity vital to our physical and emotional health and one that, following so-called "green" agricultural practices, has the potential to create massive environmental harm. In Chapter 7, I talked about "The People of the Corn", the idea that the industrial food system's food pyramid is based almost entirely on #2 corn from Iowa. There the emphasis was on the impact of these practices on the land and water and on human health. Here let's look at meat production in this system from the spiritual aspect because what you choose to eat, especially meat and animal products, is vital to an ethical relationship with the rest of nature.

Eating food is a key characteristic of all animals and the system of "eaters" and "eatees" is nearly as old as the origins of life on this planet. This means our bodies have adapted through the reproductive success of our long line of ancestors to eat the chemical substances that make up the bodies of other organisms. We are not adapted to eat artificial food—chemicals that taste sweet but contain no calories or vegetable oils that have been chemically changed to make them solid at room temperature. Our bodies are not equipped to stay healthy with vitamins coming from a One-a-day pill rather than with vitamins in the context of their naturally occurring food stuff. Our digestive tracts cannot function properly with a diet lacking in non-digestible fiber. Here I'm making a strong pitch to stay away from the cheap, unhealthy pseudo-food produced by the industrial agriculture complex. Of all the things in life, it's vitally important to know where your food comes from.

Let's start with meat. Cows raised on land near you where their feces and urine serve to fertilize the fields are a good source of dairy products and meat—they and their bounty of food are good for you and a natural part of the farm ecosystem. As I've mentioned previously, mass production of meat and dairy products using industrial CAFOs is a recipe for food poisoning and for food that contains unnatural components, such as the pigs that are now genetically engineered to produce fish oils. If you do eat this industrially produced meat, it is important for your health to avoid eating the fat, because the toxins of pesticides are fat soluble and move easily through the food chain. One way or the other, if you want to reduce your impact on the earth and improve your health, it's important to reduce the amount of meat you eat and for the sake of the earth and to avoid wasting any of it because it takes much more land to feed people with meat than it does to feed people with grain.

Besides that, I don't think it's possible for us to be nourished physically or spiritually by the meat of animals that have lived their whole lives in pain and misery eating an unnatural diet of corn. Chickens are a great example of a wonderful species that has co-evolved with us to provide us with food and eggs in return for "chicken feed."

The other day I was sitting on my lower deck where it was cool and shady. My chicken came up to spend some time with me. This isn't the first chicken I've lived with; I had a white Leghorn named Chicken for six years. People all over the earth eat chicken so this species does a great service to humans. I first got to know a chicken when I was teaching Embryology, the study of the intricacies of how and why eggs develop into whole animals with all their parts in predictable places and at predictable sizes. This is the study of the "miracle of birth" if you will. Anyway, I had chicken eggs in an incubator to collect future nerve cells for a cell culture experiment. One of the eggs got too old and hatched so I asked the students if any of them wanted a chick. None did, so that night I stuck the chick in my pocket and stopped by the grain elevator to buy some food for it on my way home. I asked them for the smallest amount of chick starter they sold while this chick was peeping loudly from my pocket. The fellow looked at me with a funny look and asked me how many chickens I had and, when I told him one, he opened a 50 lb bag and gave me some. That was a good business decision because since then I've bought about 10 bags of chicken food from them. Because I'm a bit short on creativity, the chick became "Chicken" and Chicken became an important member of our household.

She soon outgrew the small box I kept her in when she was tiny and so I built her a cage about three feet high. I thought I knew everything about that chicken given all the information about embryology I had taught to students and then living with her for a while. But when my mother, who had grown up on a farm, saw Chicken she started to laugh and told me my chicken was a Leghorn. What I didn't know, it seems, is that this breed likes to perch in high places and they are among the feistiest of the chickens. Periodically Chicken would challenge my dog or me to a higher place in our household "pecking order." She always lost, but we still respected her.

After six years, Chicken died tragically in a pond accident and, when I told my evolution students about it, one of them offered me another chick. He said his chicks were from inbred parents so one of them might be a bit odd, but I was welcome to one if I'd like. I happily accepted his offer and, yes, the chick was a bit odd. When startled, it peeped loudly and ran in a left-hand circle. Not an effective escape mechanism, but that behavior earned our new chicken its name, Little Chicken, a left-hand version of the name "Chicken Little."

Little Chicken isn't a Leghorn and she has grown up to be a wonderful companion. In the warm months, she spends most days outside taking dust baths and scratching around for worms and, in the process, turns bugs into eggs that supply the household. Little Chicken is a good friend. She makes quiet noises and is willing to sit politely in the chair next to me when I'm relaxing. We're different species, but we can communicate at some level that speaks to the commonality we humans share with other creatures. To a chicken, happiness is a warm spot to sleep, a good nest to lay your eggs in, a dust bath, and lots of bugs and plants to pick around. If you tap the ground, they will come over right away to see if you've found something good to eat. And they communicate their happiness (and distress) with lots of kinds of sounds. Modern science disconnected to spiritual values coupled with financial greed has turned the production of chicken as a primary food source for humans into an immoral horror story.

Now in the U.S., *six billion* broiler chickens per year are produced for slaughter. They are raised in perpetual dark indoors to a weight of 3 ½ lbs in seven weeks by special breeding, the addition of Vitamin A and D along with antibiotics to their diet, and with their toes and beaks removed so they can't harm one another in the crowded conditions. These are not happy chickens but they provide a cheap source of non-red meat that is served in giant helpings all over the U.S. to make the 60 companies involved wealthy. The story associated with egg production and the production of other animal products by industry, as opposed to farmers, is similar. Let me make myself clear: eating meat is not bad. Nature works that way. Little Chicken eats bugs, for example, but the bugs aren't raised in crowded unnatural conditions. Humans have been eating chickens and their eggs for eons, an arrangement that has benefited both species, but this unhealthy mass production of chicken, eggs, and other meats is not respectful of how nature works.

The industrialized monocropping of single strains of meat- or egg- or dairy-producing animals globally is a recipe for a global disaster. Nature thrives on diversity, not uniform monocrops, especially not monocrops of animals in crowded conditions where diseases can spread easily. SARS (bird flu) and mad cow disease are only the tip of an iceberg of what can happen to us if we ignore the inevitable evolving nature of the living things around us. Under these crowded conditions with no diversity in the food animals, one evolving disease can move rapidly around the world. The pain and misery of the animals and the insecurity of the food supply are outcomes of an immoral relationship to the natural world. In our own spiritual dimension, how can our bodies be properly nourished from food that has been so horribly abused? How can I as a human face Little Chicken and not be ashamed of our collective behavior?

What we do to chickens, cows, chickadees, the soil in our fields, and the air around us, we do to ourselves. The human psyche is deeply embedded in nature and, when we behave immorally in our relationship with nature, we inevitably harm ourselves and the generations that will follow us. In my mind, the only moral way to eat meat is to ensure that the animals have been allowed to live a chicken, cow, or pig life before they are slaughtered humanely. When you think about it, dying for that purpose isn't such a bad way to go. The bottom line is we must connect the lives of the animals we eat with their deaths and with the source of the meat we eat—in that way we value that food for what it is rather than as a mysterious "given" in our lives that appears swathed in plastic wrap on Styrofoam trays.

Big Ticket Item #6: Local Food

In our biological evolution, food has always been a local phenomenon—in our long history we've always eaten only those varieties of food that grow or live nearest to our homes. That is a good way to choose food even now. Eating local food and especially your own garden vegetables nourishes our bodies in the healthiest way possible. The change of diet as the seasons change is the natural way our bodies have evolved. It's amazing how good a fresh local strawberry tastes in the spring when you haven't had any except the ones you preserved from the previous year. It's amazing how bad a strawberry tastes after it has traveled several thousand miles to your table. Freshness has a huge impact on the nutritional value and taste of food. The more local you eat and the more you choose foods that are recognizably an apple or a tomato or a carrot, the better off you will be. By eating local food, you can get to know the farmers who grow your food and know about any chemicals that have been added. Whenever possible, you are much better off eating organic food. The impact of pesticide residues in food hasn't been fully evaluated but it is known that vegetables raised with chemical fertilizers have less nutritional content than those raised with organic fertilizer. Remember the lesson of *The War with the Newts* (Chapter 7). The environmental damage from fertilizers and pesticides are well known so if you wish to tread lightly on the earth it's important to avoid promoting their use by buying foods raised that way.

Big Ticket Item #7: Clothing

According to *Time* magazine[103], the average American family spends about 4% of its income on clothing. Nowadays most of this clothing is made in China out of petroleum-based plastics. If you get it too hot, it literally melts rather than burns. This clothing is displayed in a dizzying array of jammed clothing racks in huge stores and the popular styles and colors keep changing to keep us buying new clothes. We continuously haul new clothes home and try to find a place for them in our jammed drawers and packed closets. A year later, we wouldn't be caught dead in them because the styles lend themselves more to trendiness rather than to comfort and once they're washed, they no longer are wearable. Have you gotten an article of clothing home only to discover it requires dry cleaning rather than washing? If so, that's another unnecessary bill to pay and it's also bad for the environment. Dry cleaning fluids are toxic in their production, in the air near the dry cleaners, and in the residue left in your clothes. Owning clothes that require dry cleaning turns your closet into a toxic waste spot where you really shouldn't breathe the air. Obviously, it's better to avoid buying clothes that need dry cleaning and it's better to avoid buying plastic clothes.

Fortunately there's a better way. Natural fibers like bamboo, organic cotton, and hemp are durable and vastly more comfortable than synthetic fibers. They also can be washed and pressed back into shape if needed. I have both synthetic and natural

---

103 Kiviat, Barbara. 2009. "How to Save BIGGER: Need to Sprnd Less? Here's the best way to go about it. Hint: It's not just the little things that add up". Time. May 18, 2009.

fiber clothes in my closet and I find myself wearing the natural fiber clothes preferentially because they are comfortable and designed in a timeless way. For example, I have a hemp t-shirt that is cool on hot days and warm on cold days. I don't know how it knows the difference, but it works.

Producers of some of these natural fibers have successfully bred their plants to naturally produce a variety of colors including reds, purples, and others. This means you can choose clothing that hasn't been impregnated with toxic dyes. In addition, makers of natural fiber clothes frequently have developed networks of people producing a wide range of clothing styles and colors where all the people in the clothing production system are making a living wage. Because natural fibers are just that, one of nature's bounties, there is no limit on the variety of groups who could get in the business of growing and producing unique, nature-friendly, durable, comfortable clothing. They will make a living as long as we are willing to preferentially choose the clothes they produce. And we will be healthier if we don't wrap ourselves in petroleum products that have traveled around the world several times to get to our department stores.

Here again is a familiar theme. The industrial clothing system that is mass producing clothes and marketing them to us in department stores is damaging our comfort, health, and earth's life support systems. We can protect all of these things as well as our pocketbooks if we purposefully choose natural fiber clothing from diverse networks of small producers.

Big Ticket Item #8: Buy Local

The closer to home you can buy anything, the better it is. Local producers of everything from food to clothing to lumber are all better for earth and better for you because the pollution from transportation is mitigated when you buy local. In addition, when you buy local you are also helping your neighbor make a living because buying local strengthens your local economy.

There is a nonprofit organization called BALLE (Business Alliance For Local Living Environments)[104] whose main function is to promote the development of local alliances that help small local businesses. For example, imagine the economic impact of going out to eat at a local restaurant built by a local Green Built developer who has used materials that come from within a 150-mile-radius of the restaurant. Imagine further a menu with fresh, local produce, dairy products, and meat, all organically produced with care for the local soil and water supplies. The chef has prepared the food fresh with herbs grown in the restaurant's kitchen garden. Doesn't that make your mouth water? And doesn't that make you feel secure in eating that meal? And doesn't the cost you pay ensure your local economy is strong so you can be sure of a job?

Contrast that scenario with a eating at a fast food restaurant (or a restaurant chain) or doing your shopping at an international chain of department stores. In each case, only low-paying service jobs are present for your local economy and the profits are sucked out of your local economy to go into the pockets of the wealthy few who own the supply systems associated with the businesses. They have raped and pillaged communities and ecosystems around the world to develop products you are willing to buy. And remember, the goal of any huge corporation is to make ever increasing profits even if it damages their customers (Chapter 2). If you buy local, you have forced the environmental impacts and any health impacts to be immediately apparent to

---

104 http://www.livingeconomies.org/

you because they will be happening near you where you can influence change. That's how nature functions—we will be much better off if we apply the local first principle to all our purchases.

Big Ticket Item #9: Recreation

Lastly, let's talk about recreational activities. For short trips under about 500 miles, avoid plane travel because it is one of the most energy-intensive modes of travel. The CO2 and other pollutants produced by planes is released high in the atmosphere where it has more than four times the impact of pollutants released nearer the earth. Instead, try a train or an energy-efficient vehicle. You might also want to explore the vacation opportunities close to your home—that will help boost your local economy and make you more aware of the history and features of your region. Recreational activities that do the most harm to the environment are off-road vehicle travel and motorized toys of any type that use a two-cycle engine. It's better to cross-country ski than to snowmobile and better to swim than to use a jet ski. Notice in both cases how much more exercise you get by avoiding the motorized toy. If you can't live without the noise and speed and fun of motorized toys, it would be best to choose electric motors for your toy. If they aren't on the market, modify one to electricity. If you love the sound of a motor revving, you can always record a bunch of that noise and play it through your iPod while you're driving.

Of course, the U.S. is a big country and the world is a worthwhile place to see. Each of us must travel on planes to far-flung places occasionally to visit relatives or to see a new part of the world. That's ok. The airline industry has been working to decrease the pollution caused by air travel. For example, many airlines are now cleaning planes with soap and water rather than toxic cleaners. As things are now as we wait for improvements in airplane fuel efficiency, it's best to choose daytime flights during the warmer seasons of the year because night flights, especially when it is cold, produce about ten times the environmental impact of daytime, warm weather flights.

Purposeful Design in Life

This chapter could go on but I hope you see the key philosophical point—moving your life closer to the efficiency and elegance of nature will provide you multiple benefits and also protect earth's life support systems. Planning on the front end of decisions and sizing your choices not merely to your budget but to your actual needs will make your life simpler and fuller. I suspect none of us will get to the end of our lives and wish for more time working or mowing the lawn. Paul Hawken[105], writing about how businesses could change their practices to become sustainable said, "In any endeavor, good design resides in two principles. First, it changes the least number of elements to achieve the greatest result. Second, it removes stress from a system rather than adding it." These are good words to remember when you set about simplifying your life.

Analyzing carefully using nature's operational principles and a time, species, global perspective inevitably will help you see ways you can remove stress from your life and free yourself from unnecessary expenditures. This kind of thinking will also help you move closer to being a part of nature, the natural place for humans. As you do this, you will become more like the

---

[105] Paul Hawken, "The Ecology of Commerce," 1993, HarperBusiness, a division of HarperCollins Publishers, pg 166

hummingbirds, "eating" your fill while your activities promote health and abundance for the future. Of course, you can always track your success in living a purposeful life by checking your trash periodically. If you've been successful, you shouldn't have much trash at all. This, I think, is something worthy of celebration when our life inevitably ends.

Important Concepts of this Chapter

A time, species, global goal for living is to live in such a way that you maximize the comfort and fullness of your own life while generating a positive impact on the non-human living things around yourself.

Rather than trying to do 1,001 things to "save the earth," it's more effective to choose wisely on big purchases that can trap us into lifestyles that stress our budgets, impede our ability to live a full life, and decimate the lives and homes of the other species around us.

The size, construction quality, and location of our home is the most important environmental and personal decision we make. Done carefully and with a time, species, global ethical framework, we free our time, our earning power, and our lives from slavery to a disorganized, wasteful, expensive lifestyle.

Keeping the toxins out of our homes improves our own health, the health of ecosystems near our home, and health of systems near where the toxins are manufactured.

Eating fresh, local food cooked from scratch provides nutrients we are adapted to eat and increases the diversity of food organisms around the globe. Ideally, individuals grow at least part of their own fruits and vegetables in their own yard.

Eating meat is important to the health of ecosystems, but only if it is locally raised and the animal is allowed to live its life naturally before slaughter—this improves the fertility of soil and spiritually ties us to the lives and deaths of our food animals.

In all instances, waste is wrong—waste of energy, waste of clean water, waste of food, waste of our time and energy.

Freedom comes from sizing our work lives, our home lives, and our consumption to our needs rather than to a fickle, never-decreasing well of wants.

* * *

There is hope for the future because across the U.S. and in places around the globe, people are beginning to see the urgency of our need for dramatic change in our relationship with each other, the earth, and with the future. In the next chapter, we'll be looking at the possibility of widespread change and some of the innovators who have already been working on it.

Chapter 11

There Is No Hope—How Wrong Can You Be?

As we've seen in *The War with the Newts*, the reality of the world we live in is driven by economic forces. Fortunately for all of us, there is a small but rapidly growing segment of the business community around the world who are thinking green with a time, species, global perspective. These people have discovered that by working toward community vitality, environmental health, and corporate profit simultaneously, their profits increase. This group gives me more hope for the future than any other facet of human activities occurring today.

Loss of Hope

As my children got older and as I gained control of my teaching and research in southern Colorado, I decided to renew my excitement with my work by expanding my research program from a singular focus on ancient sediments and paleoclimate to include some work on the amphibians living at high altitude in the Rocky Mountains. The best candidates were boreal toads, large black amphibians with big golden eyes that breed in the snow melt water at 9,000-12,000 feet elevation. These toads have a very short active season each year because the areas where they live are cold and under twenty or more feet of snow for most of each year. My first challenge was to find a population of boreal toads. I had seen hundreds of them when I first moved to Colorado in the early 1970's, when I had collected a few for the bone and specimen collections at Michigan State University at the request of my PhD sponsor, the late J. Alan Holman. Fifteen years later, it wasn't so easy to find these exceptional animals.

Local people kept telling me that boreal toads were abundant in one spot or another so I spent each weekend walking in the mountains in one majestic spot after another without finding any toads. Finally, I started contacting people at the University of Colorado and at the Colorado Department of Natural Resources only to find my negative data were valuable as another piece of evidence in the application for Threatened and Endangered status for the toads. I kept looking for them, but became more and more emotionally depressed because they were simply gone—thousands of little deaths that no one noticed.

For at least fifteen years I had lived in the rural, isolated, sparsely-populated San Luis Valley, an area with few people partly because of its extreme temperatures—the record low temperature was 53°F below zero and only about two months each summer were frost-free. Periodically I had heard about environmental concerns but they were always peripheral to my life in this protected enclave. I had no idea that the impact of human activities had become so enormous as to affect even isolated places like the high elevation southern Rocky Mountains. Finding these exceptional toads going extinct under my nose caused my world to crash. If I lay on the ground, I felt as though I could hear the earth crying out with pain. As a biologist, earth and its plants and animals had seemed secure constants in my life—I had spent years of my schooling learning the scientific names and geographic distributions of animals around the world—all of a sudden everything changed. I literally was mourning the death of earth's abundance. To me at the time it felt like unrestrained human population growth, especially in the poor countries of the world, seemed to be killing off the other species of the world. Later I learned the cause wasn't so simplistic.

Remember Ehrlich's IPAT equation from Chapter 1? Environmental impact is magnified by the number of people, but the technology used and the resource intensity from affluence are multipliers that increase the problems. Obviously it wasn't just the fault of poor people scratching out a living in Third World Countries causing the death of the toads in the mountains around the San Luis Valley. Still, it's easier to scapegoat the least powerful than to accept personal responsibility. One way or the other, from that early beginning I have spent many years developing the ideas I have been presenting in this book.

After my deep depression, eventually I started getting my act together. I decided to use my time and talent to become an agent of change to help solve the various problems leading to the loss of boreal toads. In short, I got over my emotional depression and started working toward change, new cheese so to speak. A while later I started noticing college students with similar problems. Ignorance in environmental matters is bliss, a choice many people make, and those who open themselves to a realization of the depth of our problems can experience emotional depression and a loss of hope for the future. Ideally, action follows the shock of awakening to the reality of our times because doing nothing accomplishes a great deal—it nets you more results like the ones that are already a problem. And as John Corey, a character in the novel *Plum Island* by Nelson DeMille says "It occurred to me that the problem with doing nothing is not knowing when you're finished."

I firmly believe these system problems I've been writing about in this book are solvable if we use evidence-based decision making and harness the creativity and productivity of at least

20% of U.S. citizens. Why 20%? Because 20% are going to oppose any change of any sort and 60% aren't paying attention. Twenty percent united on a single front can change the whole system because the 20% opposed each have their own little agenda. Twenty percent of us together form a selective force that changes the evolution of whole nations—that's what it took to bring the Berlin Wall down. These ideas are parallel to a rule of thumb—the 80/20 rule—used by economists, marketers, and infectious disease specialists at the Centers for Disease Control. In their data, it appears that 80% of the work—spreading a disease, buying a product, committing crimes—is done by 20% of the group involved[106]. Examples of the 80/20 rule are all around us as viral marketing (e.g. name-branded clothing) that has become so popular nowadays and email campaigns that ask you to forward email to everyone in your address book. Similarly, a few people in each community can change whole communities to be organized as though they understand a time, species, global perspective and behave within nature's operational rules. It will take each person like those who have read this book to become active though and we must be purposeful in our efforts. To be effective, our 20% must be careful how we use our energy.

Government Follows, Not Leads

Besides modifying our own behaviors (chapter 10), the first place most of us look to for action is the various levels of our government in the U.S. and to the United Nations for international problems, both reasonable expectations given the traditional roles of governmental entities in protecting its citizenry.

---

[106] Gladwell, Malcolm. 2002. "The Tipping Point: How Little Things Can Make A big Difference". Back Bay Books/Little, Brown and Company, New York, NY.

But any reasonable person today will decide in short order that this is a losing proposition. Yes, we have the EPA and various environmental regulatory agencies in our states, but let's look at these entities in another way. As we've seen, the EPA does set an upper limit to how much pollution can be spewed into our air or water and whether wetlands can be drained or not, but these limits are political and economic decisions rather than decisions associated with what's best for people or other species. In a sense, the EPA's regulations set the amount of damage industries can do to us without fear of legal liability. In other words, EPA air or water pollution limits are "get out of jail free" zones for polluting industries even if their pollution is damaging the health of people or other species or even killing people.

I had always thought of government as an entity that protected it citizens from harm—now I know that isn't true. Gus Speth[107], co-founder of the Natural Resources Defense Council and CEO of the UN Development Programme, says that we cannot look to government to solve environmental problems because there is an almost seamless link between economic interests and the positions governments take in negotiations. In other words, industry agendas have become the agenda of our government. Evidence for this can be seen in the series of international agreements that to our shame the U.S. has refused to ratify. They include the Convention on the Rights of the Child, the Convention on the Elimination of All Forms of Discrimination Against Women, the Land Mine Convention, the Convention on Biological Diversity, the Law of the Sea Treaty, the Kyoto Protocol, threats to pull out of the 1972 Anti-Ballistic Missile Treaty, a rejection of a proposed enforcement measure

---

107 James Gustave Speth, "Red Sky at Morning," 2004, Yale University Press

for the Biological Weapons Convention, and rejection of the International Criminal Court. As it is currently run, the federal government will not step up to solve our problems and it will not cooperate with other nations to help solve global problems.

Certainly, this is an area we as citizens of the U.S. and of the world should start putting pressure but, to be honest, any change will take too long to avert disaster in the environmental, social, and spiritual crisis that is already upon us. Fortunately, there are other, more effective solutions that will lead to the changes we need in government. Biological evolution provides the mental model for the type of change needed in this situation.

Punctuated Change

When most people think of biological evolution, they automatically think about what biologists call gradualism—slow steady change that gradually modifies one species into another. For example, it's known that birds evolved from small, bipedal dinosaurs—many of the most ancient bird fossils found in museum collections were catalogued among the dinosaurs because the birds had been fossilized without feather impressions or wings. With a gradualist model of evolution, the transition from small dinosaurs to birds would entail a whole series of transitional forms with incipient feathers and a thousand stages of the modification of hands into wings. This is not what is found in the fossil record. Instead there are bursts of change with stasis in between, a pattern that didn't make much sense until biologists started to understand the program of development of animal bodies.

Changes in shape can occur in bursts by *changing the control* over pattern development in embryos. To understand this, let's go back to chickens for a minute. Chickens are used in the study of embryos because they develop in an egg enclosed by a shell that is incubated outside of the mother's body. It's easy to cut a small hole in an egg, manipulate the embryo, and then reseal the egg with a small piece of thin glass and bit of wax. The thousand stages of modification of hands into wings can be demonstrated in reverse in one generation by moving a small piece of tissue in the developing wing bud from the back to the front. The chick will hatch with hands rather than with the backward-bent, seemingly crippled "hand" that is the wing of a bird. Mutations (changes) like this are the fodder for abrupt shifts in the progress of biological evolution. The pattern is "punctuated equilibrium" rather than gradualism. With a punctuated equilibrium pattern of evolution, innovations arise in small populations and become fine-tuned there. In a relatively short period of geologic time, the modification spreads and replaces the original form. Such change causes a domino effect in other co-evolving parts of the ecosystem.

In the complex adaptive system of our culture and economy, change should be even faster than in biological systems because cultural evolution is Lamarckian—that is, we don't have to wait for genes to spread from one generation to the next or to wait for embryos to develop and grow up—change in culture can occur by lateral learning and mimicry. In other words, a punctuated pattern is characteristic of cultural evolution because of how easily replicating, competitive ideas/technologies can spread. For example, think how quickly our use and dependence upon the Internet has spread. Or think how quickly adjustable rate mortgages spread to unsuspecting consumers. Rapid, punctuated change is the hallmark of cultural change.

This means we have the ability and opportunity to quickly shift our industrialized society within our lifetimes from its old, destructive version to one that is sustainable. With this in mind, let's go back to regaining hope for the future of our industrialized civilization—we're looking for changes (mutations) that have the ability to modify the industrial base upon which our economy and well being depend. These innovations continue to be developed and perfected in small, protected spots and some are now ready to spread and replace the old industrial framework. Others are still in the development phase, but there is promise all around us.

First let's start with some of the philosophical giants of our time who have laid the mental foundation for the kind of change we need—Janine Benyus; William McDonough; and Paul Hawken, Hunter Lovins, and Amory Lovins. Together these innovative thinkers have laid down a framework for us to follow.

Biomimicry

Janine Benyus is a small woman with a big voice and a brilliant mind. Her book, *Biomimicry: Innovation Inspired by Nature*, coined not only a word but also brought attention to a way of thinking. Before Janine Benyus, innovators solved human problems in human ways rather than looking to nature as a model, measure, and mentor. For example, typical industrial manufacturing uses the adage "heat, beat, and treat" to make materials or parts. Exceptionally high temperatures are used to melt things or to cause toxic chemical reactions plus strong forces are used to mold and shape things. The whole process is poisonous, noisy, and energy inefficient. Nature, on the other hand, creates the hardest materials known, the most flexible materials known, underwater glues, and all sorts of other useful materials at biological temperatures in the presence of life. It does this using energy collected locally from the sun. Nature is so good at what it does because nature's innovations have been subjected to 3.8 billion years of evolution—nature's solutions to problems are tried and tested in the great free marketplace of the ecosystem that is ruled by the "invisible hand" of natural selection. In nature's marketplace, failures don't go bankrupt and then reorganize to foist another failed product on the ecosystem—they go extinct and their genes are lost to succeeding generations. Because the selective process is harsh, the solutions are elegant, subtle, and effective.

A case in point is the shape of fans. Most man-made fans make cavitation noise, a hum associated with the fan anytime it is turning. Typical commercial buildings have sealed windows and an engineered air handling system with the air driven by fans. This means you are subjected to this low-level background noise any time you are in a commercial building. This is an irritation and it also is a waste of energy because it takes (wastes) energy to produce the cavitation noise. One day a designer was walking with Janine Benyus by the sea shore and noticed the shape of sea shells, a shape that he made into a fan that moves without cavitation noise or wasted energy. Why on earth aren't we using this shape for all of our fan-shaped things like ship propellers, wind generators, and window fans? Similar stories can be told in materials science, food systems, energy systems, and other aspects of human needs. Janine Benyus has stimulated a plethora of ideas[108] but the key one of importance here is the concept that the more we look to nature for solutions to the vast array of challenges we have, the more we will find tried and tested solutions that work.

Cradle to Cradle

The second philosophical giant is William McDonough,[109] whose reverse intentional design assignment we read about

---

108 http://www.biomimicryinstitute.org/

109 http://www.mcdonough.com/

in Chapter 4. Bill McDonough is a unique individual partly because of his vast life experience—as a child he experienced life from the poverty of the streets of Taiwan and everything in between to the wealth of people who own a private island. He is an architect and designer, the founder of McDonough and Partners, a company that designs buildings and communities, and co-founder of McDonough Braungart Design Chemistry, a company that produces toxin-free chemicals and certifies supply-chain safety from toxins. There are a lot of stories that could be told here but I'll share only two with you.

The first is about a day care center McDonough and Partners designed—the goal was to make this a safe place for small children. One day it dawned on McDonough that toddlers chew everything they come in contact with—this led to the realization that the materials commonly used in paints, fabrics, and plastics are toxic to children. All manufacturers get their feedstock from a limited number of chemical supply companies like DOW Chemical, and the exact composition of materials is proprietary information. To give you perspective on chemical industry's power and influence, you should know this is a $1.5 trillion/year industry. In other words, in the industrial marketplace there was no way to buy chair cloth that was safe for children to lick or chew.

So McDonough along with his German chemist friend, Michael Braungart, found textile and dye companies who would help them develop a product line that was toxin-free. This was almost impossible but their breakthrough came when CIBA-Geigy, a Swiss and Japanese dye company had "hit the wall," pinched between tightening Swiss water regulations and their supply-sources for chemicals to make dyes. This company opened their books to test dye chemicals for toxicity—they found only 16 compounds out of the 1,600 commonly used to be safe. From those 16 compounds, the company was able to dye cloth any color. This move accomplished a number of things simultaneously—CIBA-Geigy decreased their regulatory costs, kept their workers safe, water leaving their plant was cleaner than the water entering it, and their profits went up. With dye from CIBA-Geigy and toxin-free cloth from a Steelcase subsidiary, McDonough was able to design his child care center with toxin-free materials. The move toward toxin-free materials in textiles has grown from there. Now Ford Motor Company is using a toxin-free seat cover made by Interface Corporation in its new cars. The essential philosophical lesson in this story is the incorporation of nature's operational rule of Waste as Food[110] into industrial applications.

Obviously industrial processes still must have some toxic components, at least for the time being. To deal with this problem, McDonough and his collaborators have developed C2C (cradle-to-cradle) thinking, the concept that any waste should be confined to a technical food chain (toxins that are essential but must never be released to the biological world) or used in a biological food chain (waste essentially becomes compost to increase the fertility of the earth). If this cradle-to-cradle thinking became the norm, there would be no need for EPA regulation at all—rather than paying environmental regulators to determine the cost/benefit analysis of health risks versus pollution limits, we could spend our time and treasure on actually solving problems!

The second Bill McDonough story is about Herman Miller's "Greenhouse." Herman Miller is an office furniture manufacturer in Zeeland, Michigan, with a strong commitment to sustainable business practices. Sustainable business uses people,

---

110 http://www.mcdonough.com/principles.pdf

environment, and profit as their "triple bottom line" rather than the old saw of "business exists only to make as much profit as possible regardless of the harm to the community, the employees, and the environment." Herman Miller contracted McDonough and Partners to design a central manufacturing facility for them. McDonough believes that humans as biological organisms need light and fresh air in order to be happy and healthy (what a surprise!), so he designed a 300,000 ft^2 manufacturing facility for Herman Miller in which the air is completely replaced every 30 minutes and there are 60 skylights. They have a "main street" where employees can buy food or eat their sack lunch and socialize. The building cost was $52/ft^2, about 15% more than comparable manufacturing space. Its energy bill is 30% less than comparable commercial space so the extra cost of construction is paid for *every three months* by the reduced cost of energy. Any business person knows this is an outstanding ROI (return on investment). There are more benefits yet to the company.

Over half the employees have 100% attendance records, the health of the employees is much better than is normal for manufacturing plant employees, and Herman Miller employees wait in line for a transfer to the Greenhouse. Now Herman Miller is working on making its products as environmentally friendly as their manufacturing facility. This is just one story among many of established corporations that have moved from a profit motive to a triple bottom line and found more profits there than with the old model. McDonough's philosophy and vision has been an international beacon that has helped countless others to migrate to a "New Industrial Revolution."

Natural Capital and Profit

Last on my list of philosophical giants are Hunter and Amory Lovins and Paul Hawken, all of the Rocky Mountain Institute[111] in Snowmass, Colorado. This story starts with an energy policy article published in 1976 by Amory Lovins in the prestigious journal *Foreign Affairs*. In the article, *Soft Energy Paths: Toward a Durable Peace*, Lovins, a physicist, argued that the U.S. should transition from a dependence on fossil fuel and nuclear power (the hard path) to an energy supply system based on increased efficiency coupled with dispersed renewable sources powered by the sun (the soft path). He advocated matching the energy need to the energy source—rather than using high-level energy sources like fossil fuel for low-level energy uses like heating spaces or water, he advocated using the laws of thermodynamics to our advantage. It's important to note the timing of this article—the U.S. had just passed its domestic peak oil supply and had started to increasingly rely on foreign oil imports. Lovins foresaw the importance of local energy sources in national security as well as the international conflicts that would arise over foreign energy sources. What a different spot we would be in now had we just listened to his insightful advice!

Since then, Amory Lovins and his wife Hunter have started and grown the Rocky Mountain Institute into a $12 million/year operation that funds itself through its own enterprise plus grant funds. Paul Hawken is a vital part of the team at RMI. These professionals are masters of thinking outside of the strong boxes of the old industrial revolution. One way they make money for the Institute is to make the business case for what

---

111 http://www.rmi.org/sitepages/pid23.php

they call "Natural Capitalism"[112]. According to Hawken and the two Lovins, the first industrial revolution led to industrial capitalism, a system that uses up financial capital, manufactured capital, and human capital. Industrial capitalism doesn't obey its own accounting principles because it liquidates capital and calls it income and it assigns no value to natural resources, living systems, or to social and cultural systems. The underlying philosophy of industrial capitalism is that humans are separate from nature: we live by our own rules and are not vulnerable to the rules of nature, ideas I hope you have discarded after reading Chapters 4 and 5. In this old system no toxin is too dangerous to use and the EPA is there only to "control the damage" so the system doesn't kill too many of us. Natural capitalism is a gigantic leap in human understanding and behavior that is needed to avoid global upheaval and breakdown and to avoid chaos from the disintegration of the finicky economic, industrial, social, and political structures built since WWII using industrial capitalism. As Amory Lovins puts it "Invention is the sudden cessation of stupidity."

Lovins' invention is the mental shift from being above and apart from nature to recognizing that we must become benevolent and beneficial citizens of earth in all of our activities if we are to persist for the long term as an important species. In particular, especially manufacturing, commerce, and all aspects of business must make this shift because the economy is so important to human well being. In our mental shift, we must think of our economy as a subset of life and living on this planet rather than thinking of the environment and our culture

---

112 Paul Hawken, Amory Lovins, L. Hunter Lovins, "Natural Capitalism: Creating the Next Industrial Revolution," 1999, Little Brown

as wholly-owned-subsidiaries of the economy. That mind-set and perspective is clearly enunciated in the quote below from the Rocky Mountain Institute[113].

- *The environment is not a minor factor of production but rather is 'an envelope containing, provisioning, and sustaining the entire economy'.*
- *The limiting factor to future economic development is the availability and functionality of natural capital, in particular, life-supporting services that have no substitutes and currently have no market value.*
- *Misconceived or badly designed business systems, population growth, and wasteful patterns of consumption are the primary causes of the loss of natural capital, and all three must be addressed to achieve a sustainable economy.*
- *Future economic progress can best take place in democratic, market-based systems of production and distribution in which all forms of capital are full valued, including human, manufactured, financial, and natural capital.*
- *One of the keys to the most beneficial employment of people, money, and the environment is radical increases in resource productivity.*
- *Human welfare is best served by improving the quality and flow of desired services delivered, rather than by merely increasing the total dollar flow.*
- *Economic and environmental sustainability depends on redressing global inequities of income and material well-being.*
- *The best long-term environment for commerce is provided by true democratic systems of governance that are based on the needs of people rather than business.*

---

113 Hawken, Paul, Amoy Lovins, and L. Hunter Lovins. 1999. "Natural Capitalism: Creating the Next Industrial Revolution". Little, Brown, and Company.

The quotation above highlights many of the things we've been discussing but the key difference here is the emphasis on the power and influence inherent in the business enterprise to shape its business model on an extractive, destructive, domination path versus on a path of collaboration, health, sustainability, and abundance. Remember how business and economic interests shape government policy? Businesses that strive to function in line with the Rocky Mountain Institute core values above behave in ways vastly different from their industrial capitalism counterparts.

With a natural capitalism framework, a business could reassign their environmental regulations manager from monitoring and influencing new EPA regulations to developing new methods to ensure zero emission of pollutants and zero unused waste from the business's activities. EPA regulations would no longer matter to them because their emissions would be so far below EPA limits that unlawful pollution events would not be a concern. Instead, the business would be focused on how to turn what had previously been waste (think of pollution as wasted resources) into another profit stream. Similarly, all aspects of the business's activities would be put under scrutiny to turn waste into food, to diversify their profit streams by capturing and using or selling their waste, and to maximize collaborations with the communities in which the business functions. At a political level, natural capitalism businesses would be supporting metrics that would accurately measure progress such as GPI instead of GDP (Chapter 3). They would also support international social arrangements that foster fair play and diversity rather than WTO or World Bank monocrops of agricultural and industrial systems that maximize their profits but disrupt the diversity and stability of countries around the world (Chapter 9).

Notice the radical change in mind-set and values shown above—this natural capitalism thinking would lead us to a system of global evolution that would move us toward security and stability for all of earth's inhabitants. With this kind of time, species, global thinking, humans and our modern civilization in particular can persist for the long term instead of becoming one of those failed civilizations Jared Diamond writes so eloquently about[114]. To my way of thinking, it is immoral to continue extracting wealth from countries around the world while ignoring the natural capital present in each community of each country. That behavior is leading to the collapse of our modern way of life and the ecosystems of the earth. Moral behavior at this moment in time must include positive action toward sustainability using a time, species, global perspective and nature's operational principles.

Viral Marketing of the New Thinking

These visionary messages of Janine Benyus, William McDonough, and the people of the Rocky Mountain Institute haven't gone unheard. Internationally and in the U.S., there is a growing cadre of corporations whose leadership is choosing the moral path to the future. There are also individual innovators and venture capitalists that are investing their time, talent, and treasure in developing and bringing new practices and technologies to the market. They are all part of a movement Gus Speth[115] calls "JAZZ"—localized outpouring of initiative from all

---

[114] Jared Diamond, "Collapse: how Societies Choose to Fail or Succeed," 2005, Viking

[115] James Gustave Speth, "Red Sky at Morning: America and the Crisis of the Global Environment," 2004, Yale University Press

segments of our society—business, NGOs, foundations, groups of private citizens, and individuals. JAZZ is systemic change from below. It has the power to quickly transform our civilization from a global destructive force to one that will lead the world toward stability and prosperity, punctuated change working in our best interest. Our role as citizens is to preferentially build our careers working for these organizations, preferentially buy products produced in sustainable ways, and help elect officials at all levels of government who "get it." We must then insist our elected representatives put people before corporate interests in their decisions. Let's take a look at some examples of the corporate pioneers of the New Industrial Revolution.

Corporate Pioneers

One of my favorites is a story about a building design described by a speaker from the Rocky Mountain Institute. His main point was that nibbling at the edges of energy efficiency for commercial buildings results in the high cost of green buildings. To decrease energy use, conventional thinking has little confidence in natural systems so architects design "green buildings" with redundant systems for heating, cooling, and ventilation. This results in buildings that would cost ~40% more in construction to save ~10% of the building's energy use. No businessman in his right mind would make such a choice, no matter how moral it was perceived to be because there is no real ROI (return on investment) due to the increased cost of redundant systems and the minor returns in energy savings. For example, you must build a complete conventional air conditioning system along with some lip service to a natural cooling system if your target is saving 10% on your energy usage. But this is where McDonough's design question is important.

Is it your intention to green wash your business or do you intend to follow nature's rules using resources only as needed, adapting your facility to its local climate, and reducing or reusing any waste involved? If you want to save only a little energy, the redundant systems are the only thing that works. If, on the other hand, the building is designed to save 60-80% of its expected energy use, then the cost of the building goes down to normal costs per square foot. Buildings of this sort are patterned after the engineering feats of termites in sub-Saharan Africa, where termite nests maintain a constant 70°F temperature in the face of 120°F heat outside of the nest. The termites make use of ventilation, air flow, and natural conduction to maintain their comfortable home. Innovative ideas like windows that open to allow workers to breathe fresh air and admit a breeze to their work space and ideas like allowing natural light into even the interiors of buildings save energy and make the commercial building a place people like to work. I don't know about you, but I've spent many years working in buildings with artificial light, sealed windows, the hum of fans to move air, and an air handling system that doles out only enough sanitized air to keep the employees (students in my case) conscious. It seems to me we would all do better if we would design buildings as though humans were living organisms rather than "production units". And we would all do better if our buildings followed nature's operational principles instead of wasting resources to make ourselves miserable in these engineered environments.

Another slant on natural capitalism can be seen in the activities of Interface, Inc. Paul Hawken's book, "The Ecology of Commerce" was the stimulus that prompted Interface founder and chairman, Ray C. Anderson[116], to dramatically re-make his

---

116 http://www.interfaceinc.com/who/founder.html

carpet company into what it is today, a firm moving toward a model of natural capitalism. Anderson realized that carpet is a petroleum-intensive product that does harm to people and the environment through its entire lifecycle. From when the petroleum is extracted from the earth, made into chemical feedstock, manufactured into carpet and dyed, installed, and then disposed of in massive piles in landfills, carpet destroys natural capital and leaves a swath of pollution and waste behind. Over a 14 year period, Anderson and his team of employees at Interface have had a "Mission Zero" to eliminate any negative impacts of their activities by the year 2020. They have made amazing progress—Interface employees are some of the happiest and most enthusiastic people I have ever met. Here's a bit of what they've come up with.

People and corporations buy carpet because of how clean new carpet feels in a space—it makes a place feel warmer and quieter plus it provides color and a sense of elegance. Purchasers aren't particularly interested in having to clear all their furniture out in order to install new carpet, plus keeping carpet clean and stain-free is expensive. Interface's business model allows its customers to buy the aspects of carpet they want while minimizing the fuss and bother. They do this by making carpet squares with carefully designed patterns out of completely recyclable, toxin-free materials—the carpet is made out of polylactic acid (from corn) rather than petroleum feed stock and it is dyed with McDonough and Braungart's non-toxic dyes. Interface doesn't sell carpet anymore; instead, they contract with businesses to meet the needs formerly met by buying and installing new carpet. Interface now provides a service of keeping a new carpet feel in commercial places. When coffee is spilled in one spot on the carpet, Interface can come in during the night and replace the damaged square of carpet. Matching color isn't a problem because the pattern in the squares mitigates any differences in dye lots of carpet. The old square is returned to Interface to be manufactured into a new square. Nothing ends up in a land fill and any way you look at the cost, it's less expensive than repeated replacement of hundreds of yards of nearly new carpet just because the traffic paths are worn. Interface now manufactures their carpet in many different styles for use in homes also.

Another innovative company is Cascade Engineering[117] headed by Fred Keller. Cascade Engineering manufactures a variety of plastic molded products like the large, rolling recycling and waste containers commonly in use around the nation. In the triple bottom line model, Cascade Engineering has focused primarily on the people piece of "people, environment, and profit." Fred Keller was deeply moved by a book called *A Framework for Understanding Poverty* by Ruby Payne[118].

Ruby Payne's life experience has, like William McDonough's, taken her from living in poverty to living among the super wealthy. From this, she developed a deep understanding of poverty as "the extent to which an individual does without resources." Most of us equate poverty with lack of money, but the resources Ruby Payne talks about go much deeper. They include financial resources but also emotional, mental, spiritual, physical (health and mobility), support systems, relationships/role models, and knowledge of hidden rules (unspoken cues and habits)—all resources vital to each person's success. Fred Keller has taken this framework for poverty to heart and is working though his

---

117 http://www.cascadeng.com/

118 Ruby K. Payne, Ph.D, "A Framework for Understanding Poverty," 1996, aha! Process, Inc.

corporation to heal the problems, one person at a time. Cascade Engineering hires the inner city poor and ex-cons. Rather than let them fail at their job because of the lack of non-monetary resources, he also hires social workers to help provide the non-monetary resources needed to keep people on the job. In addition, the climate of the workplace is designed to break the cycle of poverty by mentoring employees to address all the aspects of poverty. I suspect there are few corporations with so many totally loyal employees—people whose lives have been re-made because they were fortunate enough to be hired by Cascade Engineering. Fred Keller and his team's welfare-to-work program have been recognized as a national model program[119].

Now Cascade Engineering has a new product, inexpensive wind generators that quietly generate a portion on the electricity needed on individual homes. When I last looked in the spring of 2009, the cost of buying and installing one of these generators is about $7,000. Along with some of the new tax rebates for installing renewable energy, the savings from this sort of investment would allow homeowners to save and invest what they otherwise would pay to energy companies to burn more coal that exacerbates global warming. If you bounce the benefits of these products off the Rocky Mountain Institute mind-set principles of behavior for corporations, you can see clearly a new set of values being used in Cascade Engineering's business practices. In each place they do business, people, the environment, and their profit all benefit.

Corporate pioneers come in all sizes and with different motivations. Guy Bazzani is a big Italian man with a small but growing business in green development. There is nothing of

---

119 http://www.cascadeng.com/press/pr/20010221.htm

green wash in his work—he is constantly thinking and bringing his innovative ideas to fruition to the benefit of each community in which he works. Bazzani and Associates, Inc.[120] specializes in rebuilding trashed-out places in city centers. Guy was a developer in California specializing in building redwood decks. On the weekends, he liked to hike in the towering redwood forests that used to span the coast of California. One weekend Guy got to his favorite forest only to find it had been completely logged in his absence. Redwood is a brittle wood so the loggers pile up the forest litter to form a cushion for the mighty trees as they fall—that means any small trees are destroyed along with the large ones—essentially leaving no chance for the forest to regenerate itself. Because Guy is a thinking man, he could see how his business was causing a permanent loss of a beautiful ecosystem. He decided to turn his life's work into something that helps communities, both human and non-human. In the process, he migrated to Michigan and put his considerable talent and energy into rebuilding run down city neighborhoods into walkable areas with toxin-free, LEED-certified signature buildings that help communities reunite.

As a green developer, Guy recognizes the value of the old buildings that are present in any given neighborhood. Instead of tearing them down and wasting all of those building materials, Guy guts buildings saving everything of value but removing all sources of toxins. He then rebuilds using LEED certified methods and clever techniques of his own design. For example, to increase energy efficiency, he doesn't allow any material like a metal girder to span from inside to outside of a building because they essentially wick heat out of or into a building.

---

120 http://www.bazzani.com/

His core and shell design has earned LEED's highest certification. This exceptional work is not accomplished with all sorts of exotic supplies—all materials for his buildings come from within a small radius of his building site, usually less than 150 miles. When you walk into one of Guy's buildings, there is no "new building" smell, an indication that all of the toxins have been banned from entering the building. The cost per square foot is comparable to conventional building but the energy savings for the owners range from 47% and higher. His corporate headquarters are in an old, historical building with the offices on the first floor, his lush living quarters on the second floor, and his "yard" is a vegetative roof on the building where he and his wife Carol can hold private barbeques and parties above the smog of the city. His sense of accomplishment, joy, prosperity, and enthusiasm are infectious.

I could go on with these examples because there are many more. In the West Michigan region where most of these businesses are centered, the manufacturing downturn sweeping across the U.S. isn't hitting as hard as it is elsewhere in Michigan. There haven't been studies done yet on the specific reasons why, but I suspect the trend toward natural capitalism is a key reason. Back in 1994, the companies mentioned and nearly 100 others formed an organization called the West Michigan Sustainable Business Forum under the umbrella of the West Michigan Environmental Action Council (WMEAC). The WMSBF members meet monthly, listen to speakers, and network with other businesses striving to promote business practices that enhance environmental sustainability, economic vitality, and social responsibility. Most of what each business does is not proprietary—by sharing common values and "aha" breakthroughs, the entire group prospers and the economy and environment of the region have become more stable and healthy. This amazing group of people has grown from a few early visionaries like Bill Stough, a quiet, unassuming man whose deep commitment has changed the region to a force that is shaping the economy of the region. Bill started as a volunteer for WMEAC and his volunteerism was instrumental in giving the WMSBF a start. This is Gus Speth's JAZZ at its best.

Similar stories could be told in communities around the nation. For example, the Sustainable Business Alliance in the San Francisco Bay region has three chapters and over 250 member businesses. They also emphasize local first (BALLE) principles to strengthen their local economy first before reaching out to other places on the globe. Other examples could be given, but now you know what to look for in your own region. Companies, employers who collaborate locally to discover ways they can uniquely use the ideas of the philosophical pioneers discussed above. There is hope in JAZZ. I expect punctuated change will occur region-by-region shifting us all toward natural capitalism. Each time individual people opt to pay attention to what is occurring in the businesses of their region (or get some JAZZ started if none exists), we move closer to the 20% needed for a mind-set shift to natural capitalism.

The Grid and Alternatives

Another form of JAZZ is occurring nation-wide and globally in energy innovation. As mainstream U.S. culture is starting to realize the heavy cost of relying on imported fossil fuel for energy, a plethora of innovators and venture capitalists are working to develop a new energy system for us. This work is not trivial because our current energy and fossil fuel companies are some of the most powerful entities in the world.

Taken together, the 2006 profits of the ten largest global oil companies ($167 billion[121]) was greater than the 2007 GDP of all but 40 nations. The playing field in energy is not a free market because these powerful companies have spent decades securing subsidies, molding trade agreements and regulatory structures, and influencing governments to ensure favorable conditions for their own prosperity. Our entire infrastructure is built around what these companies say they need to deliver an uninterrupted supply of electricity, gas, and petroleum products like plastic to us. Sprawl would not be possible without unending supplies of cheap gas and massive homes with expansive turf grass lawns aren't possible without reliable sources of electricity and gas. With our current system, all of us are slaves to our monthly energy bill—think how much wealth and power you could amass if everyone paid a monthly tithe to you!

The key paradigm of our old energy supply system is mono-cropping and uniformity—central supply and central control using the same energy sources all over the country—a recipe that puts tremendous power in the hands of a few and flies in the face of nature's rules. This central supply may have been a good idea at one time, but now one snow storm can knock out electricity to several million people at one time. This is of special concern because modern houses aren't livable without a constant supply of electricity. As new technologies are developed, the key problem they face is trying to fit into "the grid" (aka "the established power structure"—pun intended) and into the old ideas about how electricity should be generated and delivered. Central generation of electricity in coal-fired plants or nuclear-

---

121 Juhasz, Antonia. 2008."The Tyranny of Oil: The world's most powerful industry—and what we must do to stop it". William Morrow, an imprint of Harper Collins Publishers.

power plants results in about 70% loss of energy between the energy of the fuel and the amount of energy available to the end user; longer distances increase the percentage lost. Much of the challenge then is to transform energy generation and use into a local phenomenon rather than a "one-size-fits-all" system. Many of the new technologies like solar and wind do best as distributed sources of energy; small versions of these types of energy collectors can be sited on individual homes. Some states even have laws that allow individual people to sell energy back to the utility companies. Wouldn't it be wonderful to have energy generation be part of your income rather than part of your monthly bills?

To meet the needs of our whole society with renewable energy though will require a substantial new grid infrastructure, partly because the best places for solar and wind are usually far from population centers. In a distributed energy system, different areas of the country and the world would make use of the type of energy most abundant in their local region and that energy could be transported using a superconductor grid that loses much less of the energy in transit. Right now these free, renewable sources of energy are valuable resources that are being wasted while we are using mainly coal and nuclear fuel for energy generation with the attendant water, air, and land pollution. Investing in Lovin's soft paths toward energy, encouraging homeowners to generate electricity, and harvesting fuel-free energy where it is abundant would be a solution to our energy needs that fits with Paul Hawken's principles that govern nature—diversity in solutions and waste as food.

To transform a system as massive as the grid and the fuel-supply system with fuel-free energy sources requires careful, stepwise change. Much of our country's wealth is invested in the old paradigm so the investments already made must be protected so investors can earn a return on their investment. At the same time, it is foolish to invest any new money in old technologies beyond basic maintenance.

Proposals to build new "clean coal" or nuclear plants boggle my mind—there is no such thing as clean coal—the technology doesn't exist yet so any attempts to build a clean coal plant are experimental. Even if such a system could be developed, the environmental damage done to recover coal from the earth is appalling. The common technique now in the U.S. is to remove entire mountain tops to mine the coal. Whole regions of West Virginia in particular have been devastated by this practice. Nuclear energy is touted as a cheap, carbon-free source of energy but it may be the worst option of all those available. Nuclear power plants can function only in the context of the grid—they must run full bore or not at all and they must have an external supply of energy to keep them from "melting down." In the summer of 2003, most of the east coast of the U.S. lost power simultaneously, an incident that cost millions (billions?) of dollars. This incident was propagated by the nuclear power plants—they go down like lined-up dominos if the grid becomes unstable—safety shutdowns to prevent nuclear meltdowns. Then there's the problem of contaminated sites and nuclear waste.

At the Michigan Sustainable Business Forum and Expo in 2004, Fred Keller of Cascade Engineering talked about a nuclear power plant in northern Michigan that was being decommissioned—he said the cost of decommissioning exceeded the value of the electricity sold by the plant in the entire duration of its existence. Since then I've spotted a charge on my electric bill for the decommissioning, an especially interesting charge because

I've used 100% green energy since I moved here and have never used the electricity from that plant. I'm sure my grandparents were supplied by it though so now I'm paying for electricity for my long deceased grandparents' TV and lights. You can see this as an especially stupid system given that electricity is ephemeral—it dissipates if not used. Why would we choose to promote a technology so costly and deadly? And why would we finance electricity for thousands of years, the time it will take for the nuclear waste to become less harmful? Still, the nuclear plants in existence should be used until they are obsolete, again to ensure a return on investment for the capital held in them. We certainly should not provide any public money to build any new nuclear or coal-fired power plants but we can make the best of the situation while we are in transition to nature's sophistication in the collection and use of energy.

In *Earth: The Sequel*[122] Fred Krupp, president of the Environmental Defense Council, and Miriam Horn argue that the most effective thing we can do to transition to a new energy system is to level the playing field of the energy market. Energy innovators aren't asking for subsidies or favored status; they simply need to have our regulatory system remove the subsidies and favored status from coal and nuclear energy. You may not be aware, but American oil companies had about $9 billion in taxpayer subsidies in 2008—this occurred even while their profits were at global record corporate highs and while gasoline prices increased to record highs in the summer of 2008. The creation of a level playing field, according to Krupp and Horn, would best be accomplished by putting a price on carbon with a

---

[122] Fred Krupp and Miriam Horn, "Earth: The Sequel. The race to reinvent energy and stop global warming," 2008, W.W. Norton and Company, Inc.

cap-and-trade system. The cap would decrease each year, causing the price of emitting carbon to increase each year. Putting a price on carbon would increase the cost of spewing global warming gases into our atmosphere and trickle down as increased expense for all polluting activities. This would, in turn, allow a window for innovators to compete in the marketplace and, as they gain market share, prices would be driven down by consumers choosing the less expensive, cleaner alternatives. The bottom line here is the idea that we must do market reform if we are to succeed in accomplishing energy reform.

Remember Paul Hawken's principle about "living on current solar income"? Many innovators are working toward developing technologies that will allow us to do just that. Solar energy can be used to generate electricity. United Solar Ovonics, for example, produces a sheet metal coated with thin layers of solar-absorbing material. This product can generate electricity even using bright moonlight. Most of United Solar Ovonics product is being bought by Germany rather than by people in Michigan; the two places are comparable in their solar power generating capability but different in their regulatory framework. Other areas of the U.S. are even better for solar power—a solar power plant is currently being built in the San Luis Valley of south-central Colorado, partly in response to the aggressive renewable energy standard recently adopted by the State of Colorado.

Solar energy can also be used to generate heat to turn turbines to generate electricity. This can be done in the sunny southwest by erecting giant parabolic surfaces to concentrate the solar energy. Such a system could be used to turn the turbines of existing coal power plants, thus eliminating the massive release of pollutants from burning coal for at least part of each day. Coupling the solar energy with wind generators could help take care of night-time energy needs because the wind blows more at night than during the day.

"The economic benefits of solar farming are also impressive. A 2006 study by engineering consultant firm Black & Veatch found that solar thermal plants create twice as many jobs as coal and gas plants and produce eight times the retained revenues in the state in which they are located. Each gigawatt of solar thermal-generated electricity, according to the National Renewable Energy Lab, will create 3,400 construction jobs, 250 permanent jobs, and $500 million in tax revenues."[123] According to Krupp and Horn, "In terms of the amount of land required, solar power plants are more efficient than wind farms, which need about six times the acreage per watt produced. Solar energy is also much more efficient than biofuels, which even with improved feedstocks and conversion processes will need more than thirty times the land per watt—although liquid fuels can do many things solar energy cannot..."

Each area has its own best source of energy. Areas like the San Luis Valley of Colorado are rich in solar energy, a fact not lost on poor local people living there. Nearly every house has a homemade solar collector to help heat homes in this sunny area where night times lows frequently are -20°F. Areas like Michigan are rich in wind potential, especially along the edges of the Great Lakes. In Michigan, wind coupled with geothermal energy could replace nearly all the domestic home heating needs for coal and nuclear power. A key point is that each local place needs to develop energy generating mechanisms adapted to that place—climate, weather, land availability, worker availability,

---

[123] Quotes from pg 65, Fred Krupp and Miriam Horn, "Earth: The Sequel. The race to reinvent energy and stop global warming," 2008, W.W. Norton and Company, Inc.

and other local considerations are vital. The most economical and environmentally sound energy source is close to home and locally adapted to that place.

Energy for Vehicles

One of my favorite stories from the Krupp and Horn book is about two innovators who embody the best of Paul Hawken's principles and who are addressing the issue of renewable supplies of liquid fuel. The farmers of the Corn Belt would like to have the liquid fuels of the future be derived from corn but that's a bad idea for many reasons. In Chapter 7, we followed a bit of the story of corn in the U.S.; here's more of the story. Corn is made into everything from paint to crayons to diapers to food in the U.S. It is a heavily subsidized crop that is one of the most damaging to our soil and water, both from the standpoint of water supply and water pollution. Unfortunately corn has already captured 90% of current U.S. biofuels subsidies. Producing corn is so energy intensive that making it into ethanol nets hardly any increased energy even while its production is damaging our ability to feed ourselves in the future and driving up current food costs. Other countries are damaging themselves as much as we are by making poor choices in their biofuel feedstock. For example, palm oil fuel is causing massive deforestation in Malaysia, and deforestation in the Amazon is being driven by turning the land over to production of soy beans and sugar cane to be made into biofuel. Fortunately there are many other ways to produce liquid fuels from biological sources.

Krupp and Horn tell an amazing story about a company called GreenFuel, an innovator by the name of Isaac Berzin, and a venture capitalist by the name of Leonard Blavatnik. Berzin searched indigenous and marine varieties of algae (the choice was purposeful to protect the local environment from invasion by a non-native species) to find one he could use as a starting point for rapid selection and adaptation to the effluents in a coal-fired plant's smokestack. They had to give the algae just the right amount of light plus a supply of water and nutrients like nitrogen and phosphorus. Their supply of water comes from a local sewerage treatment plant. So Berzin is using two streams of waste to grow algae. Once he has a full crop of algae, an event that occurs daily, he dries it and extracts a high quality liquid fuel from the algae. This whole system is amazingly brilliant and mimics nature's sophistication—using current solar energy, Berzin is able to produce clean water and remove CO_2 from the air while making a product that can be used to run trucks and cars.

With a price on carbon emissions, GreenFuel might be hired to put an algae greenhouse near the 1,700 U.S. power plants that have the space for the greenhouses needed plus on-site refineries for the biofuel extracted from the algae. This type of system could turn some of our most polluting power plants into energy generators that solve several problems at one time. The efficiency of the GreenFuel system exceeds that of the ethanol from corn or soybeans by so many factors that it is laughable we taxpayers have been forced to subsidize the wrong technology. Possibly the key lesson to be learned from this comedy of errors is the role of government should be restricted to ensuring free markets rather than to set up markets to benefit favored technologies. Because the waste heat and energy at our current utilities plants could be used to power the algae-growing operations, the GreenFuel story illustrates another trend that is occurring in the New Industrial Revolution, industrial ecology.

Industrial Ecology

In Kalundborg, Denmark[124], the industries in the city have developed a system that follows the waste as food principle. The Kalundborg industrial system is called industrial ecology or industrial symbiosis (living together). Their energy power station provides waste steam to heat the city buildings, and its waste warm water is used to run a fish farm. The fly ash from the smoke stack is used to make cement, nickel, and vanadium. The list of symbiotic relationships could go on, but I'm sure you get the idea—their corporations are linked to one another and produce products from pigs to wallboard while simultaneously reducing their waste and the cost of their feedstock by using a neighboring company's waste. Together the companies of Kalundborg save 45,000 tons of oil, 15,000 tons of coal, and 600,000 m^2 of water each year. Their CO_2 emissions are reduced by 175,000 tons/year and sulfur dioxide, which causes acid rain, is reduced by 11,200 tons/year. Waste is turned into a resource at a rate of 4,500 tons of sulfur/year, 90,000 tons of calcium sulfate/year, and 130,000 tons of fly ash/year. Notice they have turned their pollution (waste) into resources they can sell for a profit.

No Need for Green Wash—There's More Profit in Green

As Amory Lovins said, "Invention is the sudden cessation of stupidity." With thousands of innovators like those mentioned here working to solve many of our economic, social, and environmental problems, how can we lose? All we have to do is to be smart enough and caring enough to pick the true members

---

124 http://www.symbiosis.dk/

of the New Industrial Revolution out from the others who have "green washed" their dirty, damaging activities because they know people care about the future of the earth. Yes, the system has a lot of inertia and, yes, there are powerful interests driving our civilization, culture, economy, and government in negative ways. But if we pay attention and use our purchasing power and personal talents to push our system toward revitalization and renewal of human and non-human nature, then we humans can prosper along side of and inside of nature. We can choose to leave a wake of health and abundance behind our activities rather than the slick of abandoned dead zones that have marked our past.

Important Concepts of this Chapter

It's easy to feel depressed and lose hope when we think about the severity of our social, economic, and environmental problems. But doing nothing is guaranteed failure.

Biological evolution proceeds in jumps punctuated by periods of stasis. The pace of biological change is limited by generational change. Cultural evolution can also change in a punctuated pattern at a pace much more rapid than biological evolution because cultural evolution is Lamarckian.

Twenty percent of people thinking together will create change because 60% aren't paying attention and the remaining 20% aren't organized behind a single goal.

Innovations that can be adapted to a variety of local applications are occurring in isolated spots around the globe. This systemic change is sufficient to change whole governments that, in turn, can set conditions that allow innovations to compete fairly in the marketplace.

Consumer products are now starting to be made with principles like biomimicry, cradle-to-cradle thinking, and triple bottom line values in corporations. Those working in these kinds of settings are filled with hope for the future.

Our sustained economic and environmental well-being depend on changing our energy system from centralized to decentralized. If we are to "live on current solar income," we must use clever, sophisticated means to collect, use, and store energy that are adapted to each local place.

Industries can save energy, money, and increase their profit margins by siting their facilities in groupings so the waste of their neighboring industry becomes the feedstock of their own industry.

* * *

All of these chapters lead us back to the beginning. As we look at the world with an expanded time, species, global worldview, we must take on a whole new level of responsibility as citizens of this place and time. Children of the future will judge our actions—we are the "deciders" of how this great cosmic story will unfold in the future.

Chapter 12

The Great Unfolding Story

According to John Haught[125], this cosmos that is our home is engaged in a great, unfinished story. Through science we can get glimpses of the possible origins and details of that story as it has unfolded in the past. We can also study the mechanisms occurring now that keep novelty and innovation entering this great system that is our universe and our home here on earth. The details of the future of this story are in our hands now though.

At this small point in time, we humans are an important part of this unfolding story of our universe, but it's vital we realize we are a *part* of the cosmic story, not its pinnacle. We are one species among many, a special species now able to look back on and understand at some level the majesty and depth of this great work of art that is the universe and the history of the earth. There can be no greater moral imperative for us than to recognize ourselves as a strong selective force with far-reaching impacts in the complex adaptive systems surrounding us.

Big neon signs all around us are flashing a clear message about our deadly power to push these evolving systems toward destructive trends instead of toward the evolved stability reached repeatedly in this inherently changing universe.

---

125 John F. Haught, "God After Darwin: A Theology of Evolution," 2000, Westview Press

Maintaining ignorance in the face of these warning signs is unethical. Remember Hem and Haw and Sniff and Scurry? Sitting at "old cheese" even while it is getting obnoxiously stale and building up toxins that will kill our children is wrong. At this point in the unfolding story of our planet and our history as a species, we must recognize our role as selective agents, not only biologically but also in our civilization, our culture, and our resulting economy.

With an expansive time, species, global worldview and a clear understanding of nature's operational principles, we can stage a careful, purposeful revolt from the paradigms of the last 150 years. As a species we are learning more each day and we can no longer pretend to ignore the fact that our actions cause ecological collapse, poverty, biodiversity loss, and other problems around the world. We can no longer pretend to ignore our I, me, now choices are causing economic and environmental disaster even while they impoverish us spiritually.

It is critical for our spiritual lives to see ourselves as an important part of this great story. What we choose in our everyday lives, what we buy, how we nurture our children, what policies we allow our government to adopt, behaviors we will accept from our corporations, the place we give the economy in our priority system, and the standards we require of our elected officials are all selective pressures on the multitude of complex adaptive systems that surround us. We ignore that fact to the peril of our children and to the peril of the millions of species that surround and cohabit with us. The future has so much to offer that we should not throw away the opportunity to ensure health and abundance for ourselves and the individuals of all species who will live here after us.

Imagine humans with a moral system that recognizes our ability to function as a selective agent on earth. Imagine our collective moral system respects this ancient place and functions with an ever-increasing understanding. Imagine also a human spirituality that deeply respects our unique place as observers who learn, understand, and record earth's story. I hope you see with me a future of creative abundance comparable to earth's history of creative abundance. This short period of widespread use of Industrial Capitalism will become a "blip on the radar of time," a failed experiment as a mainstream human civilization but one that has produced some good innovations that can be useful in our evolving future. Personally, I can get excited about this sort of thinking.

Instead of "now" and "humans" and U.S. culture as being the sole point of 13.7 billion years of universe history, we can take our place as a special species with a special culture at a special point in time with the power to work with whatever story or plan is unfolding here on earth. Heady concept, isn't it?

Death has a special place in this plan. Here I'm talking about death of all sorts of things—myself, outmoded cultural ideas, archaic species, defunct economic systems, overconsumption, and many others. If you think about the broad process of evolution, the story could not unfold without the death of the old. We humans tend to rail at death but it is death that allows change to occur in any system. Individuals of any species are born, live, and die. Corporations do the same. Ideas and civilizations too. Building death into the system on earth is the only way to allow innovations to change the future. I don't know about you, but my plan is to make the most of my time on earth and then to join the "well of souls" after I die. While I'm alive, I'm trying my darnedest to help promote a future for the earth that includes humans.

I'll end with a story from a wonderful book by Mitch Albon. Mitch Albon is a sports personality in Michigan who spent time with a professor of his as that professor was dying. *Tuesdays with Morrie*[126] is a testament to the creative power of looking at death in a positive way. It illustrates creative innovation and C2C thinking at its best. On page 141 of my copy, Mitch Albon puts Morrie's ideas as follows: "In the South American rainforest, there is a tribe called the Desana, who see the world as a fixed quantity of energy that flows between all creatures. Every birth must therefore engender a death, and every death brings forth another birth. This way, the energy of the world remains complete. When they hunt for food, the Desana know that the animals they kill will leave a hole in the spiritual well. But that hole will be filled, they believe, by the souls of the Desana hunters when they die. Were there no men dying, there would be no birds or fish being born. I like this idea. Morrie likes it, too. The closer he gets to good-bye, the more he seems to feel we are all creatures in the same forest. What we take, we must replenish. 'It's only fair,' he [Morrie] says."

---

126 Mitch Albon, "Tuesdays with Morrie," 1997, Doubleday

Acknowledgments

I have never understood why so many authors thank their editor first in their acknowledgements and make statements about how their book wouldn't be possible without their editor. Now I understand. After asking close friends to read the first draft of my first general audience book I hired Elaine Eldridge as my editor because most of my friends either quit talking to me, had nothing constructive to say, or didn't read it at all. I can't blame them. Elaine imposed an organization on my writing, asked the hard questions dealing with logic and consistency, and helped put me in touch with my target audience. Thank you, Elaine, for being a supportive friend as well as a consummate professional.

The ideas in Thinking Green developed over many years of dealing with the searching questions of my university students in General Biology, Genetics, Evolution, and Environmental Ethics. I owe a deep sense of gratitude to each student who listened, questioned, and learned throughout my many years of university teaching.

A special thank you to Connie Ingham, who doggedly read and commented on the first version of the manuscript. Her background in philosophy, history, and thoughtful living helped dramatically to clarify my thinking on some difficult issues. Thanks also to Kayem Dunn, Nick Kokx, Mary Jo Kattelman, Jean Kokx, Rachel Hood, and John Verhagen, all of whom helped me gain insight into my writing. Special thanks to my sister, Mary Jo Kattelman, who drew Thinking Green's cover image. My Mother, Father, the rest of my siblings, and my children have all been instrumental in my thinking and motivation to undertake a project of this sort. My goal is for our genetic heritage to persist for the long term along with yours, dear reader.

Finally, I owe a debt of gratitude to the lizards, birds, snakes, frogs, worms, plants, and other creatures I share my home with. Thank you all for your generous companionship and for enhancing my well being.

While I acknowledge support and help, I keep responsibility for any errors or omissions.

Author Biography

Activist and environmental researcher Karel Rogers earned her doctorate in Vertebrate Paleontology from Michigan State University. She has held administrative and teaching positions at four colleges and universities. With over thirty years of experience in genetics, evolution, embryology, general biology, environmental science, ethics, and computer science, she is currently Professor of Biology Emeritus of Grand Valley State University and the author of a series of peer-reviewed scientific articles with topics ranging from paleoclimate reconstruction to factors causing the decline of modern amphibians. She is a past president of the board of directors of the West Michigan Environmental Action Council and a frequent guest speaker on topics like human population growth, energy, and global warming. She attended the 1992 Earth Summit in Rio de Janeiro, Brazil, and her work has taken her to Madagascar and other remote areas.

Made in the USA
Charleston, SC
01 February 2011